Artificial Intelligence, Intellectual Property, Cyber Risk and Robotics

Artificial Intelligence (AI) is the most rapidly developing technology in the current Digital Age, but it is also the least defined, understood and adequately explained technological advance. This book brings together a group of leading experts who assess different aspects of AI from different disciplinary perspectives. The book argues that robots are not living systems but the creations of humans who must ultimately be accountable for the actions of the robots that they have invented. Robots do not have ownership entitlement. The book uses Intellectual Property Rights cases, evidence from roboticists, cybersecurity experts, Patent Court judges, technology officers, climate change scientists, economists, physicists and those from the legal profession to demonstrate that while AI can have very beneficial uses for many aspects of human economy and society, robots are not living systems autonomous from human decision making. This book will be useful to those in banking and insurance, cybersecurity, lawyers, judges, technology officers, economists, scientist inventors, computer scientists, large and small companies and post-graduate students.

Ruth Taplin (PhD London School of Economics and Political Science [LSE] and Graduate Diploma in Law) is Director of the Centre for Japanese and East Asian Studies, London, Editor/Founder of the *Interdisciplinary Journal of Economics and Business Law* (IJEBL), a Routledge Featured Author and has written/edited 24 books and more than 200 articles.

Routledge Studies in the Growth Economies of Asia

Artificial Intelligence, Intellectual Property, Cyber Risk and Robotics

A New Digital Age

Edited by Ruth Taplin

Routledge
Taylor & Francis Group

LONDON AND NEW YORK

First published 2023
by Routledge
4 Park Square, Milton Park, Abingdon, Oxon OX14 4RN

and by Routledge
605 Third Avenue, New York, NY 10158

Routledge is an imprint of the Taylor & Francis Group, an informa business

British Library Cataloguing-in-Publication Data
A catalogue record for this book is available from the British Library

ISBN: 978-0-367-85754-7 (hbk)
ISBN: 978-1-032-41887-2 (pbk)
ISBN: 978-0-367-85756-1 (ebk)

DOI: 10.4324/9780367857561

Typeset in Times New Roman
by Newgen Publishing UK

Contents

Preface

A new digital revolution is occurring in the emergence of Artificial Intelligence (AI) which is basically the ability of AI/Learning machines to process masses of (big) data more rapidly than humans, finding patterns from data input by people that are useful for humans to predict, create models and understand more accurately natural phenomena through AI machine vision. AI can have both negative and positive effects on all aspects of human life and nature in global terms. Yet it is a deeply misunderstood technology with humans creating frightening scenarios of killer robots and AI robots taking control of the world. This is disinformation which AI can assist in blocking through mitigating cyberattack, but it cannot do this independently. The proof that AI is not a living system, which again is a human construct, is shown through cases involving Intellectual Property Rights (IPR) which in recent times have seen attempts to claim that robots can be inventors/creators that should be awarded patent rights. In this book we submit that robots are not living systems but the creations of humans who must ultimately be accountable for the actions of the robots that they have invented. Robots do not have ownership entitlement. We use IPR cases, and evidence from roboticists, Patent Court judges, technology officers, climate change scientists, economists, physicists and those from the legal profession to demonstrate that while AI can have very beneficial uses for many aspects of human economy and society, robots are not living systems autonomous from human decision making.

This book will be useful to those in banking and insurance, lawyers, cybersecurity personnel, judges, technology officers, economists, scientists, inventors, computer scientists, large and small companies and postgraduate students.

Acknowledgements

The Editor and author of four chapters in this book would like to thank the expert contributors to this volume for their considerable efforts with their chapters.

Others who have made invaluable contributions are:

Lord Justice Colin Birss, of the Court of Appeal with his precise and astute knowledge of patents and what constitutes inventor status.

Takashi Ikegami, Professor at the Department of General Systems Sciences at the University of Tokyo, Japan, for his understanding of what constitutes a living system and for writing the Foreword.

James Brewer, former Insurance Editor of *Lloyd's List* for invaluable insurance and Cyber Risk expertise.

Peter Sowden, Routledge Editor, for his continuing support for more than 20 years.

Illustrations

Figures

Table

Abbreviations

AAI	Automobile Association Ireland
AGI	Artificial General Intelligence
AI	Artificial Intelligence
AL	Artificial Life
API	Application Programmatic Interface
CAI	Cognitive Artificial Intelligence
CBCGDF	China Biodiversity Conservation and Green Development Foundation
CBE	Classical behavioural economics
CIO	Chief Information Officer
CITES	Convention on International Trade in Endangered Species of Wild Fauna and Flora
CMMI	Capability Maturity Model Integration
CoE	Centre of Excellence
COVID-19	Coronavirus disease 2019
DDoS	Distributed denial of service
DNS	Domain Name System
DSA	Digital Services Act (EU)
EU	European Union
GDP	Gross domestic product
HPS	Human Problem Solving
HR	Human resources
IEEE	Institute of Electrical and Electronics Engineers
IOCR	Intelligent Optical Character Recognition
IoT	Internet of Things
IP	Intellectual property
IPA	Intelligent Process Automation
IPCC	Intergovernmental Panel on Climate Change
IPO	Intellectual Property Office
IPR	Intellectual Property Rights
IRPA	Intelligent Robotic Process Automation
IT	Information technology
IUCN	International Union for Conservation of Nature

LSE	London School of Economics and Political Science
MBE	Modern behavioural economics
MIT	Massachusetts Institute of Technology
ML	Machine learning
NHS	National Health Service (UK)
NLP	Natural Language Processing
NOAA	National Oceanic and Atmospheric Administration
OCR	Optical Character Recognition
OKRs	Objectives and Key Results
PDF	Portable Document Format
ROI	Return on investment
ROM	Robotic Operating Model
RPA	Robotic Process Automation
SaaS	Software as a Service
SARS	Severe acute respiratory syndrome
SDM	Service Delivery Management
SFGA	State Forestry and Grassland Administration (China)
TCM	Traditional Chinese medicine
UAV	Unmanned Aerial Vehicle
UK	United Kingdom
UN	United Nations
UNFAO	Food and Agriculture Organization of the United Nations
USA	United States of America
US	United States
WEF	World Economic Forum
WHO	World Health Organization
WRI	World Resource Institute

Contributors

Amelia Yuen Shan Au-Yeung is Associate Professor in Strategy and International Business and Interim Dean at the Claude Littner Business School, University of West London. Her research focuses on innovation, retail and internationalisation strategies in emerging markets such as China.

Victor Bartenev graduated from the Moscow Institute of Physics and Technology with a PhD in Biophysics in 1981. From the early 2000s, he began independent research on the natural science foundations of the global economy, united under the name 'Physical Macroeconomics', publishing a book and a number of articles related to this interdisciplinary topic.

Kenneth Friedman has published in physics, philosophy and economics and is Book Review Editor for the *Interdisciplinary Journal of Economics and Business Law*. He teaches at Regis University, Colorado, USA and is the President of a small publicly traded minerals company. He holds a doctorate in the Philosophy of Science and an MS in Physics from the Massachusetts Institute of Technology (MIT), an MA in Philosophy from Harvard University and under an American Council of Learned Societies Fellowship studied in Brussels with Ilya Prigogine who won the Nobel Prize in chemistry for his work on the development of non-linear thermodynamics.

Alojzy Z. Nowak is Rector of the University of Warsaw, Poland. As an inter-disciplinary economist with a PhD from the Warsaw School of Economics, he has written/edited a number of books and many articles/book chapters.

Ruth Taplin holds a PhD in Economics from the London School of Economics and Political Science (LSE) and a Graduate Diploma in Law. She is the Founder/Editor of the *Interdisciplinary Journal of Economics and Business Law* (IJEBL) and Director of the Centre for Japanese and East Asian Studies, London. As the first Routledge Featured Author in Asian Studies, she has written/edited 24 books and more than 200 articles in respected, refereed journals and has written many book chapters. She was a consultant to the Federation of Electronics Industry, UK for nine years and specialises in Intellectual Property Rights, especially in East Asia.

Chin-Bun Tse obtained his PhD in Finance from the University of Manchester and an MSc in Information Technology from Queen Mary University of London. His recent research topics include how the capital markets react to companies' announcements and disclosures, which can be examined by Natural Language Processing (NLP). Previously, this was done manually and investment opportunities were often missed because of the slow manual process. He is Research Professor in Finance and Accounting at the Claude Littner Business School, University of West London.

K. Vela Velupillai is a retired academic, living in Stockholm, Sweden. He was educated at the universities of Kyoto, Lund and Cambridge for the BE, MSocSc and PhD degrees, studying under Nicholas Kaldor and Richard Goodwin at the University of Cambridge for the latter. He has held tenured Fellowships and Professorships in the UK, the United States, Italy, Ireland, India and Mexico and published books with leading publishers and articles in respected, refereed journals. He has published with Routledge for many years. Emeritus Professor Velupillai is the founder of the field of Computable Economics.

Paul Whiteside is a Fellow of the British Computer Society, a consultant, and Chief Technology Officer with a 25-year career in technology at successful start-ups, scale-outs and large global companies. He has helped many businesses to effectively implement and scale technology, and deliver high-value, technology-enabled business change. He currently advises private equity investors on technology acquisitions, growth and value creation strategies for their portfolio companies. He is currently a Senior Practitioner at Crosslake Technologies, London, UK.

Foreword

How to Make Living Machines?

Is Artificial Intelligence (AI) a serious threat to humanity? This question has been debated by Stephen Hawking, Niklas Boström and many other luminaries. However, no matter how much we debate this question, we cannot foresee whether AI will ultimately become a great threat to mankind. In the case of atomic bombs, responsibility lies more with the humans who use them than with the bombs themselves. But the marvel of AI represents a crisis that may exist independently of humans. It is free will and autonomous decision making that has not yet been realised in AI. But it is this autonomy that will determine the future of AI, and whether it will be a threat or a boon to humanity in a new sense.

Although not as well-known as AI, my research is in the field of Artificial Life (AL). In a nutshell, AL is an attempt to make birds by looking at birds, rather than making planes by looking at birds. Airplanes are convenient, they have significantly accelerated human interaction and have brought about economic growth. AI has facilitated this. However, airplanes can still fail, albeit less often these days. Even now, when I fly, I am sometimes a bit fearful. But what about birds, by comparison? A bird – that is, a living plane – might end up flying to San Diego on a whim when it was supposed to go to Los Angeles. This is very inconvenient for someone travelling in a hurry. Rushing to work is not for such living planes. That is why AI is for a busy society. But a bird, which is a living plane, will rarely fall. There is no doubt that birds are much safer vehicles when compared with airplanes. Living airplanes might even move more rapidly if encouraged to do so.

This is because, unlike today's airplanes, the most important requirement for a bird is its own survival and relationship with other living creatures. In other words, to create a living machine, that is what AL is all about. There is a sense of trustfulness, unlike AI. Life does not die of its own accord.

In other words, the fear that AI will become autonomous – that is, that it will act freely and free from human control, that it will take away people's jobs, and that it will eventually drive humanity out of existence – is not justified. AL is directed first towards oneself, towards self-preservation and homeostasis, but it does not attack people. As will be introduced in Ruth's book,

I am now developing a living robot. A living robot must have a mind installed, and the mind emanates through interaction with people.

Today's robots are yet to become life. Why did Professor Rodney Brooks ask in *Nature* in 2001[1]: Are we missing something? Perhaps the model for building a robot is too simple? Maybe the computational power is not good enough? Or do we require a fundamental theory, after quantum mechanics and chaos, that we don't know about yet? What is the last drop, the last drop that is necessary for AI to become life? I call it Brooks' Juice, after Professor Brooks, and AL is a journey to find this Brooks' Juice.

What we have come up with is the contagion of the mind. If it is possible to transmit the mind from human to robot, it will become a chain from robot to robot. The mind is contagious through change, just as a newborn child first imitates its mother, but after a certain point, the mind changes from imitation to its own unique movement, a process I have termed an offloaded mind. Consciousness that is like a person's is contagious from person to person. That's what it takes to build a living machine. Michael Tomasello,[2] a developmental psychologist, asked the question, "Do apes ape?" Apes can imitate other's behaviour, but they do not ape the style of the behaviour. They imitate the goal of the action, not the process. Imitation of process is not a question of what to do, but how to do it. Yet, apes raised by people are different.

When we think of the fear of AI, there is the fear of not being able to see the mechanism of the machine. We are afraid of the black box. We don't know what it will do. But we are not generally afraid of human beings, who through capricious behaviour are a much worse black box. No, actually, even humans are still afraid of people they have never met before. But as we live together, we become less afraid. Paradoxically, people love and cherish other people because they are black boxes.

Maybe that's the point. An AI where we can predict the action cannot be a friend. It becomes a slave. That is because it cannot love. An AI with a heart will be loved, and the world will move towards a new state of symbiosis with different living machines.

There is a difference between a mind that does not know what it is and a black box that simply does not know how it works. In fact, the difference may be subtle. This book can help you to understand the difference.

Professor Takashi Ikegami
Professor at the Department of General Systems Sciences
at the University of Tokyo, Japan

Notes

1 'The Relationship between Matter and Life', Rodney Brooks, *Nature*, Vol. 409 (2001), pp. 409–411.
2 'Do Apes Ape?', M. Tomasello, in C. M. Heynes and B. G. Galef, Jr (eds), *Social Learning in Animals: The Roots of Culture* (Academic Press, 1996), pp. 319–346.

1 Artificial Intelligence, Intellectual Property, Cyber Risk and Robotics

An Overview

Ruth Taplin

Artificial Intelligence (AI) is a relatively recent technology although its roots lie with Dr Alan Turing's work which will be covered in Chapter Two of this book. It has developed rapidly in the last few years and is a deeply misunderstood technology. We therefore attempt to clarify the nature of AI, explaining its roots, how it can exacerbate Cyber Risk, that it is underpinned by Intellectual Property Rights (IPR) concerns and is ushering in the new age of robotics. The tendency to intensify Cyber Risk is again shown in the banking and finance sector. Related to my book *Managing Cyber Risk in the Finance Sector*,[1] intense cyberattacks first occurred in the banking and finance sectors as this sector represents both money and power which both criminal actors and geopolitical adversaries seek in their quest for self-gain. The same is occurring with the use and misuse of AI. It is furthering the Cyber Risks found in the banking and finance sector, shipping, energy and corporate sectors both large and small. Furthermore, like cyberattack, AI Cyber Risk is also becoming part of geopolitical gain between rival state actors. Therefore, it is imperative that governments and individuals understand the strength and weaknesses of the new AI technology in both furthering and diminishing state and global conflict. Yet, the hyperbole surrounding AI as a destructive force is exaggerated, with AI being just as capable of delivering efficient and effective solutions to boring repetitive tasks which make too many office jobs tedious, supporting the indispensable work of creators, being integral to the mitigation of climate change, stopping the hacking into the weak link of supply chains and freeing people's time to be more creative.

Dr Turing's work and legacy has continued to influence AI research and innovation in recent times. In March 2019, three pioneers in AI research exploring neural networks to construct living robots won the Turing Award which was initiated in 1966 and is viewed as the Nobel Prize for computing. These were Drs Geoffrey Hinton who later worked for Google, Yann LeCun for Facebook and Yoshua Bengio for IBM and Microsoft.

Neural networks are loosely based on the web of neurons found in the human brain and are a complex mathematical system that can machine learn (ML) a wide variety of sophisticated tasks through analysing and finding patterns from huge amounts of (big) data. Data can, for example, be voice

DOI: 10.4324/9780367857561-1

messages from mobiles, records of insurance claims or crop yield production trends.

In a paradigm shift from the past in which data and coding systems were input into computers manually and slowly, the learning machines could learn coding behaviour on their own at great speed beyond the capabilities of humans.

By 2010, Dr Hinton and his students had assisted Microsoft, Google and IBM in advancing both speech and image recognition. The latter was based on an algorithm that had been developed by Dr LeCun. Dr Bengio developed systems that could understand natural language and the technology to generate fake photographs that look as if they are real. All these developments will be covered in this book, but the essential question remains as to whether these new AI advancements are genuine intelligence. "We need fundamental additions to this toolbox we have created to reach machines that operate at the level of true human understanding", Dr Bengio said.[2]

The areas in which these questions of AI intelligence – and to what degree, if any, ML systems are living systems – are most debated and tested relate to whether AI robots can be awarded patents which revolve around ownership rights and what constitutes a living system.

Artificial Intelligence and Intellectual Property

Underlying all the discussions related to AI, Cyber Risk and the development of robotics is an oft-forgotten dimension of the underlying role of intellectual property (IP). We will assess all the IP-related dimensions such as AI technology being viewed as independent creators who can be granted patents, robots considered as life forms that have adopted their own autonomous intelligence, the oft-hyped idea that robots will turn on their human creators, the reality of 'killer' robots and many more related concerns.

Intellectual Property Rights (IPR) are intangible assets that are protected by law in most global jurisdictions but can be infringed. Infringement is a complicated legal judgement when it occurs between humans or via human companies. Can a robot be sued for infringement? Robots are machine learning technology and are created by humans for humans, so can they be viewed in law as an independent autonomous technology? The use of AI has ethical implications for IPR both for stakeholders and wider society in general. This is because AI algorithms are based on human opinions and biases that may be accurate or not. Therefore, AI has the potential to stereotype and may not be as objective as assumed by many. This is reflected in the ongoing firing of researchers at Google who continue to criticise built-in human biases in the development of AI at the technology giant (tech giant).[3]

Bots (AI machines) are adopting ever wider-ranging abilities which involve being efficient and forming human relationships which suggests that guidance is required on their use and application. In relation to defence applications, for example, guidance will certainly come from governmental levels, which in turn are affected by geopolitical considerations as not all governments are democratic, but can be autocratic serving the needs of a dictator and his/her acolytes. The requirements of balancing the positives and negatives in the use and application of bots are still in their infancy. Human behaviour is unpredictable, and often capricious, so the imitation game of robots learning from humans is both ad hoc and error prone.

This was reflected in a number of recent judgements globally as to whether AI learning machines in their many forms can have independent rights and be awarded patents for their inventions as autonomous creators. In July 2021, the South African Patent Office granted a patent to a robot DABUS (Device for the Autonomous Bootstrapping of Unified Sentience), for a food container invention and neural flame; while two days later, on 30 July 2021, Australian Justice Jonathan Beach of the Federal Court of Australia ruled that AI can be listed as an inventor but cannot be listed as an applicant or grantee.[4]

The breadth and divergence of global opinion was reflected in a recent court ruling in the UK in September 2021 determining whether AI machines can be awarded IPR.

The UK panel decided, by a two-to-one majority that an inventor must be a real human person under UK law. Lady Justice Elisabeth Laing wrote in her judgement, "Only a person can have rights. A machine cannot. A patent is a statutory right and it can only be granted to a person". Lord Justice Arnold agreed and wrote, "In my judgement it is clear that, upon a systematic interpretation of the 1977 Act, only a person can be an 'inventor'".

The third judge, Lord Justice Birss, took a different view. While he agreed that "machines are not persons", he concluded that the law did not demand a person to be named as the inventor at all. He wrote, "The fact that no inventor, properly so-called, can be identified simply means that there is no name which the [IPO] has to mention on the patent as the inventor". Instead, the IPO (Intellectual Property Office) "is not obliged to name anyone (or anything)".[5]

There has been much criticism of these decisions which will be discussed in Chapter Three concerning IP, AI and the granting of patents to learning machines.

This question has more gravitas than 'open access' – an attempt initiated by technology giants who dominate the internet to deny creators their acknowledgement as innovators and compensation for their talent and efforts, while taking all creative content for themselves for free. If robots, for example, are granted autonomous creator status and can therefore be granted patents, what becomes of the responsibility of the creative individual who can blame the robot for damaging behaviour without being liable for any of its actions? In subsequent chapters, we assess this question from the court case rulings, and in relation to Japanese living systems robotics expert Professor Takashi

Ikegami and robotics entrepreneur Professor Rodney Brooks; the latter is Panasonic Professor of Robotics (Emeritus) at the Massachusetts Institute of Technology and author of *Flesh and Machines: How Robots Will Change Us* (Pantheon Books, New York, 2002) among other well-known books. Professor Ikegami argues that life comes before intelligence so that once robots become intelligent, they will be concerned with survival and protecting themselves rather than destroying human beings. He has created a group of three robots named Alter who can imitate humans and each other, but questions if this is real life. Professor Brooks takes a more pragmatic perspective, viewing robots as assistants for human beings in their old age such as Roomba the vacuum cleaner robot that he created. Yet Ikegami, who is Professor at the Department of General Systems Sciences at the University of Tokyo, thinks that robots need more complexity – much greater than quantum mechanics – to develop, and ponders the question of what constitutes Brooks' Juice (his term) which is the essence of life.[6]

A New Age of Robotics?

In this book we show how nuanced the development of AI is and how we must resist the simple inevitable path of viewing it as robots taking over the world. Accordingly, we show that the roots of AI may be found in the fundamental science to understand machine learning developed by Dr Alan Turing both at the University of Cambridge and Bletchley Park where he decoded machine messages by the Nazis to make an outstanding contribution to Britain defeating the Third Reich's tyranny. AI is also rooted in the physical sciences and we note that a full understanding of the nature of AI depends on a grounding in the physics-based dynamic of thermodynamics. The contributors (especially Chapter Seven) also explore how AI will impact labour and both the positive and potentially negative aspects of job losses and the need for retraining of people for newer areas of future work such as renewable energy. Robotic Process Automation (RPA) is discussed in detail in Chapter Five which illustrates how AI robotisation can be of great assistance in automating mundane repetitive work, but can cause great problems for companies that rush to adopt RPA without assessing all the potential negative consequences. AI is also having a profound effect on geopolitics including cyberwarfare, conventional warfare, the theft of Intellectual Property Rights (IPR) and data (Chapter Three). The use of AI in the mounting rivalries for power and domination between China, Russia and the United States (US) and the counterbalancing QUAD nations of Australia, Japan, India and the US will determine the future of the economy, society and the state of the planet. This leads to the issue of AI, pandemics and climate change (Chapter Six), making this newest technology the ultimate technological determinant of the health of the planet including food security and the sustainability of regional environments such as rainforests, wetlands and other habitats to sustain wildlife.

Climate Change, AI and Robotics

Artificial Intelligence can be used to good effect to assist in climate change mitigation which is one of the most pressing problems globally. As climate change is negatively affecting the economies and societies of countries such as India, AI can be used to lower carbon emissions and provide data processing to assist in ending severe air pollution, flooding, land degradation and to help with vital drug innovations.

A recent example is using satellite tracking to understand where severe air pollution is emanating from through monitoring carbon emissions. This is particularly important for countries such as India which are trying to move away from highly polluting coal-fired power plants. Such monitoring can be used to understand other sources of air pollution in industrial and electricity plants, and the resulting data can be used to levy taxes on the heaviest polluters or to convince financial backers and the government to finance lesser polluting companies and plants. AI can also be used to reduce the time and cost of producing drugs for elderly populations and against viral infections such as COVID-19. India is a centre of generic drug production and could upgrade its pharmaceutical drug industry through AI innovations.

As noted in my last book *Cyber Risk, Intellectual Property Theft and Cyberwarfare*,[7] Cyber Risk weaknesses are often the greatest when third parties are involved such as suppliers who are targeted by hackers, criminals and state actors who enter indirectly the computer networks of large corporations to usurp, mimic or destroy the ultimate goals of the military and other government agencies. AI technology, since it can be programmed by humans to organise mundane administrative tasks which are not supervised by human quality controllers, can ostensibly allow cyberattacks to enter the systems of companies and government organisations. Yet it is the strength of AI data processing, which can find patterns in huge amounts of data, that will be the most formidable deterrent against cyberattacks which are plaguing individuals, companies of all sizes, public and private organisations as well as defence organisations.

The war in Ukraine has produced unsettling cyberattacks from China which appear to be taking advantage of instability caused by the Russian invasion of Ukraine to further Chinese espionage goals. It was reported by *The Times* on 23 February 2022, the day before the Russian invasion, that hackers allegedly from China began targeting Ukrainian websites. The hackers then targeted Russia, Belarus and Poland. Cybersecurity experts said the cyberattacks were spying on nuclear power activity, organisations and governments. Strangely, the cyberattacks were not as covert as usual but appeared under a 'false flag' so they could ostensibly be blamed on countries from the West. The United States has been concentrating under the Biden Administration on countering any possible cyberattacks from Russia and China.[8] We discuss in subsequent chapters how AI is being used and developed to mitigate often complex cyberattacks from a variety of geopolitical actors.

AI and Cyber Risk

Since AI, as argued in this book, is not a process that – contrary to much science fiction – can develop or act beyond human programming and control, room for human error continues to offer potential for mistakes and misguided action.

A report by Debevoise & Plimpton, *Tips for Creating a Sensible Cybersecurity and AI Risk Framework for Critical Vendors*[9] notes the increased risks faced from third-party vendors. Such Cyber and AI risks are at their highest when companies such as banks share confidential data with their suppliers and vendors or give the latter direct access to sensitive information. The use of AI technology can increase risk when it is organised by companies and independently by their suppliers and vendors. These linked Cyber and AI risks are being taken seriously by US Federal Reserve Agencies; on 30 October 2020 a number of them, including the Board of Governors of the Federal Reserve System, the Federal Deposit Insurance Corporation and the Office of the Comptroller of the Currency issued a Joint Paper to increase operational resilience in the banking and finance sectors. This response was much quicker given the losses banking and finance suffered before Cyber Risk was understood fully. It is positive that AI risk is being acknowledged despite the fact that AI technology is often misunderstood and can be used to solve the very problems that it is accused of creating.

This Joint Paper points to a number of actions that be taken to mitigate both Cyber and AI risk through careful risk management. This means, for example, in real terms, strengthening companies against operational risk by third-party vendors and suppliers through continuous monitoring, due diligence, negotiations and termination of contracts. In order to manage risk effectively, companies need to choose vendors and suppliers who show the least risk and analyse how much risk the company can tolerate. To do this, the Joint Paper notes that companies need to create a list of what they believe are the high-risk factors of vendors and suppliers in relation to Cyber and AI risk and decide which measures are best for reducing risk. This can involve, for example, developing contracts that address and protect against such risks.

There are a number of high-risk factors that can be isolated and mitigated. One is when vendors and suppliers have access to a company's intellectual property (IP) including trade secrets. Often the most dangerous time is when a company is working on a joint company agreement or is in the process of litigation. A second high-risk factor is when vendors and suppliers hold very sensitive information such as the details of personnel, bank statements, credit cards or medical information or have a history of leaking or having poor defences against cyberattack. Vendors and suppliers may also be deemed high risk if they use AI systems that have no human involvement, operating without specialised human quality control, as the underlying algorithms that created the programme which runs them may be breached and no one will notice.

Third, Cyber Risk, attacks and the development of AI have all become a danger to infrastructure as hackers can use AI programmes to damage infrastructure such as ships, electric grids and engage in acts of cyberwarfare.[10]

Fourth, many companies are very quick to adopt new AI technology as they believe it will save labour costs or execute a mundane task more efficiently without human error due to boredom or carelessness. A classic case of this, which will be covered in Chapter Five of this book, is Robotic Process Automation (RPA) which has been adopted by many companies in the banking and insurance sector. Lack of preparation and familiarity has led to some companies being so unfamiliar with such a novel technology that they have been forced to stop using it after mistakes were made. There have been attempts to prevent such misapplications of RPA by having those whose jobs are being replaced by RPA actually train the robots to take over their jobs. However, this training also has pitfalls as the resentment of those whose jobs will be replaced can lead to mis-training of the AI robots. If deployed quickly and on a large scale, such inputted mistakes can be difficult to rectify and costly for the company.

AI robotic programmes such as RPA when used to make decisions concerning, for example, mortgage eligibility in banking services or the extent of insurance coverage necessary to cover Cyber Risk within a comprehensive insurance policy, can cause problems for customers if they do not have access to humans. Human contact is often needed to explain why a decision has been made; how the decision can be adapted to a particular customer's requirements. Distress can be caused as customers may not understand that the decision has been made by a robotic AI process and are confused, believing that they have no human person to appeal to, especially on decisions that will place them at a disadvantage.

AI and cyber systems are so integral to the stability of the organisations that they serve, it is possible that banking, finance, insurance and medical companies that they support can collapse under the weight of both cyberattack and mishandled AI robotics.

Weakness of Supply Chains

Damage to companies' and governments' AI and cyber systems are mainly made through the weakest links and this is often through supply chains. One of the largest and most destructive known cyberattacks came through a software chain supplier, SolarWinds; the attack was discovered in December 2020 by a cybersecurity company FireEye which had also been a target of this huge cyberattack by Russian intelligence. SolarWinds is a very large American software company that had 18,000 of its clients hacked. These clients included to date 400 US electricity companies and corporations. The main target of these attacks in the United States and Europe – including, for example, the British National Health Service (NHS) – was nine American government agencies. These included the treasury, commerce, justice and state departments, nuclear laboratories, the Department of Energy, and parts of both the Pentagon and the Department of Homeland Security. It must be noted that the latter is charged with the task of keeping America safe.

SolarWinds was not the only software supplier used as a conduit into government agencies and corporations. Microsoft and other cybersecurity companies such as Malwarebytes were targeted for subsequent attacks. The United States was not the only country which had software supply chain companies breached. In France, software supply chain company Centreon had its clients' cybersecurity breached, including that of Électricité de France, the world's largest producer of nuclear energy, which holds very sensitive information. Additional companies that were breached include Orange, one of the world's largest telecommunications companies; Thales which produces precise instruments used, for example, for space exploration; Airbus, Air France and ArcelorMittal, one of the world's leading mining and steel corporations. Russia's SVR intelligence agency, known for its previous attacks on the White House and the US State Department is the most likely culprit. These cyberattacks may have started as early as 2017 under the Trump administration and were slowly seeping through the corporations, cybersecurity firms and government through low-key, back-door methods that were difficult to uncover.

The SVR, the Russian geopolitical actor behind the SolarWinds attack and others, mainly for the purpose of espionage, is not the same as the military intelligence agency known as the GRU which breached France's Centreon software supplier. The GRU hires a disparate group of Russian hackers that operate under the name of Sandworm. This group of hackers carried out deadly and crippling attacks on Ukrainian infrastructure – particularly, electric power grids in a bitterly cold winter in 2015; and then in 2017 initiated the NotPetya attack that wiped out data at Ukrainian government agencies, railways, ATMS and petrol pumps. This deadly cyberattack then spread to companies globally with crippling effects on data at a huge number of organisations.[11]

AI has the potential to deal with these very serious cyberattacks on the weak link of supply chains. As most hacking occurs through emails, an AI system, through its capacity for isolating patterns more rapidly than humans can, could be used to develop an understanding of expected behaviour across email communications. Such an AI development could be used to neutralise any deviations that constitute the cyber threat. Previous correspondence can build an understanding of whether the sender had communicated with the organisation before, whether the domain is recognised and whether a valid business relationship exists with the domain. AI can not only analyse the frequency of interactions but also the type of language used that could signify a threat.

Another indication of threat is whether links in the sender's emails have ever been accessed before by the receiver's domain or if the link is hidden from the body of the email. The AI system has the capability to find out what are both normal and abnormal patterns which it uses to protect the enterprise. From hundreds of pieces of data, AI systems are capable of determining threats and the most appropriate responses to contain the threat without aggressive offensive action that can hamper productivity and business operations.

AI systems have developed the subtlety to give only legitimate emails access to an enterprise's system without a complete blocking of all emails being sent.

Cyber attackers understand the limitations of traditional protection systems. AI has the flexibility through being able at speed, far more rapidly than any human, to process huge amounts of data, finding all the anomalies in patterns of the construction of emails that can indicate threat and respond to it quickly without disruption to the entire system of business operations. As ransomware becomes the most frequent form of cyberattack, AI systems are well-equipped to deal with such threats.[12]

Geopolitical Implications

Both China and Russia are investing heavily in AI, especially in the area of espionage, and are in advance of the United States in some new developments. However, in a surprise development of 15 December 2020, the Pentagon introduced an AI robot as a mission commander for a spy plane. In keeping with the chess game roots explained in Chapter Two of this book, algorithms used by computers to play chess and video games were used to create a software package for a robot co-pilot that shared the control of operations. An AI algorithm named Artu which was named after a *Star Wars* robotic droid, R2-D2, was in charge of the sensors and navigation of a U-2S American Dragon Lady spy plane. This reconnaissance plane was first flown by humans only in 1955 and then in 1960 was made famous by the Steven Spielberg film *Bridge of Spies*, which chronicled the shooting down of the plane piloted by Francis Gary Powers during a mission deep within Russian territory. Powers was convicted of espionage and served time in a Russian jail and labour camp until freed in an exchange of prisoners between the United States and Russia.

In a test flight in California before this historic flight of a human pilot with an AI programmed robot as co-pilot, the AI robot served as mission commander and held complete radar control over the Dragon Lady spy plane. The significance of the teaming of a human pilot and an AI robot was lauded by Will Roper, assistant secretary of the US Air Force for acquisition, technology and logistics who stated, "Putting AI safely in command of a US military system for the first time ushers in a new age of human-machine learning. Failing to realise AI's full potential will mean ceding decision advantage to our adversaries".[13]

Terms of Reference

AI is not a uniform single term for just one technological aspect of what we are assessing. It is a broad term with some common features. We will mainly be writing about different facets of AI which are outlined below and in relation to this book's chapters.

An algorithm is a mathematically-based code of rules that are input into a computer to cause a particular outcome. This is the basis of RPA (Chapter Five) which uses a set of mathematical rules to train the computer to learn how to process data such as insurance claims which has the outcome of whether the claimant is eligible or not to claim on a particular policy.

Machine learning (ML) represents a paradigm shift; it is now used by all sectors from financial to shipping to manufacturing that have moved from traditional data processing where humans are needed to give explicit instructions through programming a computer to indirect programming whereby data are fed into a computer and the computer learns to extract from the data pattern sets that provide required information. To use the insurance claim example, the data input into the computer will be processed by the computer, comparing millions of like claims to produce an answer as to whether this one particular claim can be met by the insurance company.

A Machine Learning Algorithm is one that a machine learning model learns from data and can be termed an inducer.

A Machine Learning Model is a learned programme that maps out the data inputs to the prediction stage. This can be used for a linear model or for a neural network and can be termed a predictor, as the goal of ML is to make accurate predictions based on data patterns at a much faster rate than a human data processor could do.

Interpretable ML is comprised of the models and methods that allow the predictions of ML to become understandable by humans.

A Dataset is a table of the data from which the machine learns and contains both the features and the target to predict. The dataset is termed training data when used to induce a model.

An instance, also known as a data point, consists of the feature values and the possible target outcome and is basically a row in the data.

Features are the basic inputs used for classification or prediction and is a column in the dataset. Features are assumed to be interpretable but this could be considered to be a big assumption.

The Target is basically the information that the machine learns to predict. It is an instance or target that is expressed in mathematical models.

An ML Task is a combination of a dataset with features and a target. There are a number of different targets and the Task can take the form of clustering, classification, regression, outlier detection or survival analysis.

A Black Box Model in ML describes a model that cannot be understood by looking at its parameters such as neural networks because it does not reveal its internal mechanisms. The opposite of the black box is a white box which, given interpretability, is capable of treating machine learning models as black boxes. Therefore, if, for example, one pixel is changed in one of two images that are exactly the same, the interpretation of the images can still be the same despite one being an altered image.

A Prediction is what the ML model assumes should be the target value based on the given features.[14]

Different Branches of AI

To provide a clearer understanding of AI and its different applications we provide an outline below which will make it easier to understand when reading subsequent chapters.

Machine Learning (ML) is a form of deep learning that can be both supervised and unsupervised. It can be used for complex financial models and audits. In this book it will be referred to in particular in Chapter Six on climate change as ML or reinforced learning can pinpoint patterns of certain data to make predictions. In Chapter Six, there are many references to ML assisting in predicting which regions are most prone to forest fires or when to plant crops and which crops to plant according to predictive data outcomes.

Natural Language Processing (NLP) is another aspect of AI that will be covered in this book. In Chapter Five on RPA, machine translation, text generation and answering questions are all used in, for example, robotised call centres, chat boxes for insurance clients or advertisements for accounting firms offering financial services. Sentiment analysis is part of this process which, for example, gauges the response of those watching the advertisements to see how effective they have been at persuading companies to use their services.

Expert systems that are diagnostic and make decisions are part of back-office functions which are becoming increasingly robotised. The example was given earlier in this chapter insurance claims: an insurance claim for personal injury, for example, can be made rapidly after the AI robotic process analyses the expert medical diagnosis, the viability of the claim, how the injury occurred in the circumstances outlined in the claim and how much compensation is due.

Another aspect of AI is its image recognition function through machine vision. Chapter Six on climate change has many examples of machine vision operating through satellites in space that can, for example, indicate where and how much pollution through carbon emissions is occurring in power plants globally; or image sensors through infrared cameras can see how much carbon is locked into the canopies of forests and is being released if illegal logging occurs.

Other applications of AI can be text to speech or speech to text which can assist those with visual impairments or obviate the need for a member of an executive branch of an office to have someone transcribe speech or vice versa.

Finally, the use of AI is probably perceived as most important for companies through the development of robotics which makes the work of most businesses more efficient and less labour-intensive. The latter issue of using less labour is covered in Chapter Seven, discussing the implications for workers of AI processes being introduced into businesses of any size or sector.[15]

Notes

1 Ruth Taplin (ed.) *Managing Cyber Risk in the Finance Sector: Lessons from Asia, Europe and the USA* (Abingdon; Routledge, 2016).

2 'Turing Award Won by 3 Pioneers in Artificial Intelligence', Cade Metz, *The New York Times* (27 March 2022), www.nytimes.com/2019/03/27/technology/turing-award-ai.html.

3 'Another Firing among Google's A.I. Brain Trust, and More Discord', Daisuke Wakabayashi and Cade Metz, *The New York Times* (2 May 2022), www.nytimes.com/2022/05/02/technology/google-fires-ai-researchers.html.

4 'India: South Africa Grants a Patent with an Artificial Intelligence (AI) System as an Inventor – the World's First!!', Utkarsh Patil, *Mondaq* (19 October 2021), www.mondaq.com/india/patent/1122790/south-africa-grants-a-patent-with-an-artificial-intelligence-ai-system-as-the-inventor-world39s-first.

5 See the actual judgement for all the relevant details: *Thaler* v. *Comptroller General of Patents Trade Marks And Designs* [2021] EWCA Civ 1374 (21 September 2021) (www.bailii.org) Also see 'AI Cannot Be the Inventor of a Patent, Appeals Court Rules', BBC News Online (24 September 2021), www.bbc.com/news/technology-58668534.

6 Prof. Takashi Ikegami's Webinar *Between Man and Machines*, The Daiwa Anglo-Japanese Foundation, 2 December 2021, https://dajf.org.uk/event/between-man-and-machine. Our correspondence will be referred to in Chapter Three.

7 Ruth Taplin, *Cyber Risk, Intellectual Property Theft and Cyberwarfare: Asia, Europe and the USA* (Abingdon; Routledge, 2021).

8 'Mystery of Alleged Chinese Hack on the Eve of Ukrainian invasion', Gordon Corera, BBC News Online (7 April 2022), www.bbc.com/news/technology-60983346.

9 Debevoise & Plimpton, *Tips for Creating a Sensible Cybersecurity and AI Risk Framework for Critical Vendors*, Avi Gesser, Anna Gressel, Zila Reyes Acosta-Grimes and Michael Bloom (16 February 2021), www.debevoise.com/insights/publications/2021/02/tips-for-creating-a-sensible-cybersecurity.

10 Ruth Taplin, *Cyber Risk, Intellectual Property Theft and Cyberwarfare: Asia, Europe and the USA* (Abingdon; Routledge, 2021).

11 'My Journey into the Dark World of Cyberwarfare', Nicole Perlroth, *The Times Magazine* (25 February 2021), pp. 34–37. See also Ruth Taplin, *Cyber Risk, Intellectual Theft and Cyberwarfare: Asia, Europe and the USA* (Abingdon; Routledge, 2021).

12 'Fighting Supply Chain Email Attacks with AI', Tony Jarvis, Director of Enterprise Security, Asia Pacific and Japan at Darktrace (www.darkreading.com, 28 January 2022).

13 'AI Robot Takes Control of US Spy Plane', Mike Evans, *The Times* (21 December 2020), www.thetimes.co.uk/article/robot-co-pilots-us-military-plane-in-ai-breakthrough-l83c3c0ps; 'Biden Urged to Back AI Weapons to Counter China and Russia Threats', Leo Kelion, BBC Online (1 March 2021), www.bbc.com/news/technology-56240785.

14 Christoph Molnar, *Interpretable Machine Learning: A Guide for Making Black Box Models Explainable* (2019). https://christophm.github.io/interpretable-ml-book/.

15 'Embed Artificial Intelligence to Drive Your Business', Webinar from Moore Kingston Smith that took place on 10 February 2022, www.youtube.com/watch?v= O_-pmhVc77k. Part of the Enterprise Series Webinar, used with the permission of Becky Shields, Partner, MKS. Some of the material used in this paragraph is drawn from this webinar.

2 Mechanizing Chess Games, Computable Enumerability and Dynamical Systems

K. Vela Velupillai

In this chapter, the historical lines of research by *Alan Turing*, *Claude Shannon* and *Herbert Simon*, are explored via their pioneering research on the possibilities of *Computer Chess*. Mechanizing chess and the resulting computably enumerable sets and the basins of attraction of viewing chess moves as dynamical systems, in finite game playing, makes it possible for machine intelligence and learnability to be analysed effectively.

1 Introductory Speculations

> *The game of Chess* is not merely an idle Amusement. Several very valuable qualities of *the mind*, useful in the course of human Life, are to be acquired or strengthened by it, so as to become habits, ready on all occasions. For *Life is a kind of Chess...*
>
> (Franklin, 2004, p. 317 [before 28 June 1779];
> capitals in original, italics added)

As Kawabata said, 'You do not *learn* about your opponent's *character* when you *play Go* or when you *play chess*. ... Trying to judge the opponent's character perverts the whole *spirit* of the *game*' (1972 [1951], p. 79; italics added).

Just in the twentieth century, Munshi Premchand[1] and Yasunari Kawabata in literature, Leo Tolstoy and Lewis Carroll as authors of different genres, Marcel Duchamp and Gabriel Orozco in art,[2] Satyajit Ray and Ennio Morricone[3] in films and music, the Dalai Lama and Pope John Paul II (Karol Wojtyla) representing a version of Buddhism and the Catholic Church, Vladimir Lenin[4] and Mao Zedong,[5] the one a political revolutionary and the other a guerrilla warfare strategist (and, of course, a political revolutionary), Charlie Chaplin and Humphrey Bogart as famous actors, Albert Einstein and Richard Feynman as mathematical physicists, Edmund Landau and Luitzen Brouwer[6] as mathematicians are *some* of the luminaries who were passionate about *chess*, *shogi* (Japanese Chess) or *GO*.

I feel it is possible to interpret their professional endeavours in terms of the *deep*[7] interests in one or other of these board games. This makes it possible to make sense of Kawabata's observation about the futility of trying

DOI: 10.4324/9780367857561-2

to understand the *character* of an opponent. After all, machines – whether intelligently learning or not – do not try to 'judge the opponent's character'. Machines, and humans, seem to *know* (in epistemological ways) the futility of trying to define *character* and still maintain the *fiction* that the board is made up of a *finite* number of squares – whether the game being played is chess, shogi or GO.

In computer (board) games, machines *play to win*, in whatsoever way *winning play* is defined. After all, the classic volumes of Berlekamp *et al*. (1982) is titled *Winning Ways for Your Mathematical Plays*. As Kawabata says, 'The world of Go', as a board game, may 'have its *conscience* and its *ethics*' (1972 [1951], p. 47; italics added); if so, the machine must have them, too.

This chapter is structured as follows. In the next section (2) there are some remarks on the pioneers of (digital) machine computations, particularly in relation to playing chess games. In section 3, chess play is interpreted as finite Lachlan games. Section 4 is devoted to interpreting moves in chess as dynamical systems. The concluding section (5) is mainly a plea for *a purely machine-defined conceptual space*, in which humans learn and show intelligent behaviour (whether in board games or life, in general).

2 The Pioneers – *Turing*, *Shannon* and *Simon*[8]

Michie (1986) commented that '*Game playing* was an early domain of interest, and Shannon, Turing, and Newell, Shaw and Simon contributed classic analysis of how *machines* might be *programmed to play chess*' (p. 133; italics added).[9]

I shall not unnecessarily duplicate what is widely available, in numerous articles and books, on the remarkable distinctive contributions to digital computer-based chess machines made by Alan Turing, Claude Shannon and Herbert Simon; of the many books and articles available for any interested person to consult, I myself have found most useful the two edited volumes of Levy (1988a, 1988b), Newborn (1997, especially chapter 2), but above all Hodges (1983, pp. 211–212) and Newell and Simon (1972, chapter 4); (obviously also the Turing, Shannon, Newell and Newell *et al*. chapters in Levy's books).

I would like to concentrate on a few of the remarkable insights that these pioneers have left as their respective legacies to digital computer-based chess-playing machines. Michie (1986, particularly chapters 1 and 2) are relevant for understanding the pioneers' work on games in general, but on chess machines and their algorithms, in particular, using trial and error and distinguishing puzzle vs. game learning, in finite processes.

Alan Turing

[T]he *machine* must be allowed to have contact with *human beings* in order that *it* may adapt itself to *their* standards. The *game of chess* may perhaps be rather suitable for this purpose, as the *moves* of the *machine's opponent* will automatically provide this contact.

(Turing, 1992b[10] [1947], p. 124; italics added)

In one of the early modern forays into **computer chess**, Alan Turing maintained that machines 'must be allowed to have contact with human beings' in actual situations. Adaptation *is* learning by intelligent machines and the moves in a *finite* game of chess provide the means for machines to *learn intelligently*. Turing, however, began to think of digital computer-based chess-playing machines at least six years earlier (Hodges, 1983, pp. 211–212), whilst in full flow at Bletchley Park.

However, even Turing seemed to have been mildly prejudiced in favour of human beings! He wants machines to have contact with human beings, but not the other way about: human beings having contact with machines so as to learn from them. I am sure he – as well as Shannon and Simon – would have applauded the feats of the GO-playing software *AlphaGo Zero* (Singh, 2017; Silver *et al.*, 2017), but it must be remembered that it is a purely practical solution (to which the three pioneers were not averse).

In any case, in these observations by Turing (i.e., the quotation above),[11] he distinguishes between a machine and a human being; machines are *not* surrogate human beings. Second, he does *not* consider the Kasparov alternative (Sadler & Regan, 2019, p. 8) of machine *versus* human beings, as against machine *plus* human beings.[12] Third, the set of moves, of the machine and its opponent (either a machine or a human being) form a dynamical system. Since chess is a finite game of perfect information, with moves alternating between the two players, subject to the first move being decided by some rule (tossing a fair coin, for example), it is amenable to theoretical treatment – *without any axiom* – as a *Busy Beaver Game* (Rado, 1962). Therefore, Rado's perceptive observation is relevant here:

> [W]e used in our constructions only the following 'principle of the largest element': If *E* is a non-empty, *finite* set of non-empty integers, then *E* has a largest element. ... Our examples ... show that this principle, even if applied only to *exceptionally well-defined sets E*, may take us **beyond** the realm of **constructive mathematics**.
>
> (Rado, 1962, p. 884; bold italics added)

That machine behaviour 'may take us beyond the realm of constructive mathematics' is what Turing means – I think – by 'it [the machine] may' have to adapt to the standards of human beings. For human behaviour, leading to moves in a game of chess can, *at most*, be *within* the realm of constructive mathematics.

There are three insights associated with Turing that are not generally mentioned in the context of digital computer-based chess-playing machines.[13] First of all, Turing's construction of chess machines assumed the practical workability of Hilbert's programme in strictly *finite* cases; hence, for example, the utilization of the *tertium non datur* and *proof by contradiction* was allowed in studying the behaviour of chess machines. Chess was, for Turing, a *finite game*. Second, building a digital computer-based chess machine was similar to the construction of Bombes (electro-mechanical devices) to decipher the German Enigma machine; it was a mathematical exercise based on his two fundamental contributions, the law of large numbers and the Turing Machine. Third, constructing a digital computer-based chess machine was a stepping-stone towards *machine intelligence*[14] and *learning machines.*

Turing (1992b [1947]) does allow *unorganized* machines to be organized as universal machines by means of *character-expressions* (and situation-expressions, 1992b [1947], p. 121). These expressions *can* make finite machines into infinite (countable or uncountable) machines, especially because 'character may be subject to some *random* variation' (1992b [1947], p. 121; italics added). This can be considered a fourth Turing insight and towards a theory of applying neural network theory and algorithmic information theory in the construction of machines that can play a game of chess *intelligently* and *learning* in the process. For this we must go beyond the mathematics of the traditional Turing Machine and have recourse to the Oracle Machines (or relative computation), whilst assuming that chess is a (countably) infinite game (Copeland, 1999).

Claude Shannon

According to Shannon, 'The investigation of the *chess playing problem* is intended to develop techniques that can be used for more *practical* applications' (1950, p. 657; italics added). It must be remembered that Shannon was a Professor of Communications *Science* (1957–1958) before becoming a Donner Professor of *Science* at MIT for more than 20 years (1958–1979). He was, essentially, an applied mathematician in the classic sense, not just a non-pure mathematician (although he had been a research assistant in mathematics in 1938–1940).

He met Alan Turing for the first (and 'only'?) time in 1943 (Giannini & Bowen, 2017). As Copeland and Prinz (2017, p. 344; italics added) surmise: "*Possibly* [Turing] *told* Shannon about the *computational chess* ideas that he [i.e., Turing] had previously discussed with Good and Michie, though no records exist of Turing's conversations with Shannon, so we shall *never* know for sure."

Shannon shared with Turing many of the personal traits that characterized them – both positively and negatively; for example, both appreciated 'ingenious machines and gadgets' (Ioan, 2009, p. 262), Shannon emphasizing the *engineering* aspects and Turing the mathematical and logical ones. Ioan went on (2009, p. 262; italics added): 'Chess-playing machines fascinated him; as early as 1950 he wrote a paper on *programming a computer to play chess*'.[15] Their fundamental disdain for convention, even as pure or applied mathematicians, but with immense respect for the former as providing foundations for the latter, is best described by Hodges (2008, p. 4):[16] 'It was typical for [Turing] ... to seek to outdo Bell Telephone Laboratories with his single brain and to build a better system with his own hands.'

Shannon, on the other hand, as a *scientist*, never harboured *any* doubts about the rigour of the engineering approximations of *his* constructions, nor did he need *any* confirmation on *his* 'intellectual toughness' – Turing sought the comfort of mathematically 'rigorous' approximations of *his scientific* constructions, whether it be the Turing Machine, Enigma machine, chess machines, the Riemann 'machine', or whatever.

In consonance with this observation, Shannon in his *constructions* always *assumed* that playing chess with a machine was *finite* (but theoretically, i.e., mathematically, infinity was in the background). Relay and switching circuit constructions, on the theoretical basis of (a slightly) modified propositional algebra, lay as the underlying (complex) mechanism for chess machines.[17]

In conclusion, I would like to add that Shannon was well aware of the differences between standard *Boolean Algebra* and the way propositions were derived from (even) complex circuits, the distinction and equivalences between analogue and digital computers, the truth table constructions of the calculus of propositions – hence of (Łukasiewicz) many valued logic – and the value of the Dirac delta function. He was also aware of *Boole*'s reliance on the exclusive interpretation of the OR connective (as distinct from *Jevons*'s inclusive OR[18]) in both circuit theory and propositional calculus.

Herbert Simon[19]

In 1991, Simon wrote: '*Chess* has become a standard tool in *cognitive science* and *artificial intelligence* research (a standard "organism," like Drosophila or Neurospora in genetics). Powerful programs ... . use extensive chess knowledge [but] belong to A.I., *not* to cognitive science' (p. 221; italics added).

It is interesting that Simon distinguishes artificial intelligence, AI – see, above, endnote 14 – from cognitive science. For, after all, Simon was one of those who, at Dartmouth in 1956, was enthusiastically promoting AI, mainly against cybernetics (of which he was an early advocate, see [again] Velupillai, 2018, chapter 4). At the time of his death, he was Professor of Psychology and Computer Science at Carnegie Mellon University, in Pittsburgh; with much justification it is possible to say that cognitive science is (at least) a subset of Psychology and AI that of Computer Science.

So, is Simon subject to some form of schizophrenia, at least with respect to an analysis of chess games? I think *not*!

Simon's most extensive analysis of chess is in the three chapters, 11, 12 and 13[20] of the book *Human Problem Solving* (Newell & Simon, 1972). In this book (and, of course, in his many articles, singly and jointly with others) Simon views chess (moves, positions and goals) as *Information Processing*

Systems involved with *Human Problem Solving* (HPS) underpinning *his* approach to *behavioural economics* – which I refer to as *classical behavioural economics* (CBE, contrasted with *modern behavioural economics*, MBE; see Kao & Velupillai, 2013) – of which cognitive science and artificial intelligence are parts.

Below, I view chess moves *dynamically*, in terms of the *dynamical system* approach to modelling them mathematically, but it must be observed that Simon views chess moves dynamically as IPS, *or* IPS as dynamic and chess moves as a subset of this class.

Although Simon, as much as Turing and Shannon, viewed the potentialities of the machine, of artificial intelligence and of the neural network implementation of computers, he does not seem to have embraced the *Church–Turing thesis* and the concomitant *effectivity* as the only or desirable way of *algorithmizing* HPS, within CBE (see his summary of the interactions with Kleene, as the Editor of *The Journal of Symbolic Logic*, and the difference with Hao Wang's way of proving the Theorems of Part I of the *Principia Mathematica*, in Simon, 1991, pp. 209–210). Simon has more in common with the algorithmic evolutionary economics of Nelson and Winter than with the Church–Turing thesis that underpinned the effective algorithms of Hao Wang.

My considered opinion is that Simon, more than Turing or Shannon, wanted to investigate why humans did what they did, the way they did it – and let machines do what they did, whatever the way attributed to them.

Thus, Turing and Shannon devised machines to implement the effective algorithms they – in effect, Turing Machine-implementable programs – envisaged as playing chess; Simon, on the other hand, effectivized HPS by viewing programs as an heuristic search process, operating under the Gödel *completeness theorem* for propositional algebra.

'Recent' research results of Gandy, Sieg and others cast any amount of doubt on the idea that Turing worked with the assumption of the so-called Church–Turing thesis; Shannon never assumed this thesis, especially not for the finite machines he constructed (to play chess). Thus, it is entirely possible that Turing, Shannon and Simon worked within *Hilbert's finitary program* (Novikov, 1964). I believe that in making machines to play chess, these three pioneers *constructed* machines, proved theorems about their functioning and achieved halting configurations of machines reaching well-defined goals (win, lose, draw). In this they were *all constructivists*, at least in the case of *finite* machines playing a *finite* game (such as chess).

Finally, reiterating a theme I have developed already, every mathematical economic theory must be developed for finite economies, peopled by finite agents who are, moreover, finite in number; all other aspects of production, exchange and resource constraints have respected the finiteness criteria – for example, relations (in the algebraic sense). If not, we are guilty of indulging in what Ramsey (of course, in a different context) called ethically indefensible postulates.

3 Elementary (Finite) Lachlan Games

'It seems that *human beings love games*, win or lose. It is said that in *playing a game* a *special part of the brain operates*, and I believe that *games* have a quality that *fascinates people*' (Takeuti, 2003, p. 71; italics added).

Lachlan (1970) claimed that (p. 291; italics added): '[E]very theorem of $T(\mathcal{R})$ [of recursively enumerable (r.e.) sets] *known at the present time*[21] can be proved by *constructing* an effective winning strategy for a suitable *basic game*'.[22]

I shall use a slightly modified form of the Soare (2016, §2.5, p. 43, ff)[23] definition of *finite* Lachlan games to interpret an alternate (computerized chess, GO, HEX and so on) perfect information game.

The Definition of a Finite Lachlan Game

$\omega = \{0, 1, 2, 3, \ldots\}$; i.e., set of nonnegative integers.[24]

  i Player 1 (the 'player') constructs a finite sequence of c.e. sets $\{Un\}n \in \omega$;
 ii Player 2 (the 'opponent') next, constructs a finite sequence $\{Vn\}n \in \omega$;
iii Thus, at every move of the game (of chess) the player enumerates, at most, finitely many integers and the opponent enumerates (if possible), at most, finitely many integers; therefore, at the end of stage s, a finite subset of Z_s is the result of the actions by the player and the opponent;
 iv Each play of the game ends at the end of $p \in \omega$ moves;
  v In advance of the game to be played, a finite c.e. sequence, $\{Re\}e \in \omega$ of (finite) requirements is specified precisely;
 vi The opponent wins if Z_p satisfies the precisely specified requirements – the player wins, otherwise;
vii A game *situation*, at the end of stage s is: $\{Un,s\}n < s \wedge \{Vn,s\}n < s$ and this can be *effectively* coded by a (positive) integer (cf., § 2.3, Soare, 2016);
viii A winning strategy for a player or opponent is a function from game situation to (set of) moves, coded, say, as f on ω, *s.t.*; if the player (or opponent) follows this strategy, s(h)e will win against any sequence of moves chosen by her/his opposite number.

Thus, given the above (substantiated) claim by Lachlan, to prove a theorem of c.e. sets, it is only necessary to construct a winning strategy in a win-lose, alternate, perfect information game. This is what is achieved in the final condition of the definition of a finite Lachlan game, now interpreted for chess.[25] Therefore, Lachlan's Lemma 1 can be applied, with an interpretation of σ that is deducible from the *c.e. sequence of requirements* based on all of the definition elements above for finite chess games. The slightly modified Lachlan lemma 1 for finite chess games is:

Lemma 1

For *any* winning strategy of a chess game, σ ∈ *c.e. sets*.

Remark 1

Recall that not all members of c.e. sets are computable. This remark is the rationale behind Corollary 1 (and Lemma 3).

Remark 2

Lemma 1 is particularly true of Euwe (2016 [1928]).

Conjecture 1

Every finite game of chess is an element of (finite) constructive mathematics and the c.e. sets are strict subsets of this kind of mathematics (i.e., c.e. sets ⊂ finite constructive mathematical sets).

Corollary 1

Chess is a Rado-type *Busy Beaver Game*.

Proof

Obvious, given the definitions in the Rado quotation in section 2, above and Shannon's computation for the length of a finite chess game (see Levy, 1998a, p. 6)

4 Chess *Moves* as Dynamical Systems

I start this section by quoting Stewart (1991, p. 664; italics added): 'Indeed, virtually *any* "interesting" question about dynamical systems is – in general – *undecidable*'.[26]

Simon, as discussed in the previous section, considered *chess moves* in terms of (dynamic) IPS in HPS senses in CBE. In this section I want to interpret *chess moves* as forming a dynamical system with the outcomes/ goals (win, lose or draw) as members of the basins of attraction of dynamical systems; I interpret, with Turing, *undecidability* of the attractors of the basins of attraction of dynamical systems as the *non-Halting behaviour of Turing Machine* algorithms. For this I must show that the attracting set of the basins of attractors of the dynamical system interpretation of chess moves is a c.e. set.

The chess moves of player 1, say white, forms either a *piecewise nonlinear dynamical system* or *mapping*, of (in principle) large, but finite components; similarly, for player 2, who is assumed to play the black pieces on the 8 x 8 chess board. For convenience, I shall assume that the two players' moves form, respectively, *sets of mappings and these sets are c.e. – that is, the elements that form the set are, each one, c.e., which means that the individual moves of the respective players are a finite sequence of c.e. sets. Therefore, the first four components – that is, i to iv – of the definition of a finite Lachlan game are satisfied.*

As for the *requirements* – that is, components v and vi – of the definition of a finite Lachlan game – I assume that the basins of attraction, of the moves considered as mappings, also form c.e. sets. Then, components vii and viii are trivially satisfied. Therefore:

Lemma 2

For *any* winning strategy of a chess game, with moves considered as c.e. elements of a mapping, the basins of attraction themselves form c.e. sets.

Lemma 3

The basins of attraction are *undecidable*.

Proof

Based on the observation made in *Remark 1* (above).

Remark 3

In Lemma 3, the undecidability of the attractors of the basins of attraction means, first of all, that any Turing Machine algorithm may not halt because not all elements of the c.e. sets are computable; secondly, and trivially, therefore the mapping is subject to the Halting Problem of Turing Machines.

Lemmas 1 and 2 mean that the play of a chess game can be analysed in terms of computability theory or dynamical systems theory.

5 Concluding Speculations

As Shannon memorably said: 'I am rooting for *machines*. I have always been on the *machines' side*' (quoted in Sloane & Wyner, 1993, p. xxix; italics added).

Machine intelligence[27] and learnability by machines take on a new life by means of the computably enumerable sets and dynamical system interpretations of aspects of playing chess as win/lose/draw finite games. The uncritical assumption of the distinction, or the lack of clear delineation, between the (practical) finite and the (ideal) infinite has plagued machine intelligence, learnability by machines and playing chess games. It helps to clarify the role of finiteness in playing chess games, which helps in viewing the basic limits of machine intelligence and learnability by machines.

The mathematical foundations of finiteness in playing chess games are brought out clearly in re-viewing Lachlan games as steps in the direction of constructive analysis, independent of the Church–Turing thesis. This makes it possible to implement algorithms in studying machine intelligence and learnability by machines without assuming the Church–Turing thesis – and the dimensional difficulties are highlighted by bringing in the (finite) problems of the Busy Beaver Games.

I prefer to *restart* the program of studying machine intelligence in terms of the suggestions made by House and Rado (1964), in the context of the noncomputability of finite sets, as I have done for finite chess games and the resulting c.e. sets and the basins of attraction of dynamical systems modelling (practically) of the dynamics of chess moves and the outcome space.

I also prefer to concentrate on the learnability of finite machines proving propositions without assuming axioms. I have not assumed any axioms in sections 3 and 4 of this chapter, although they seem to be part of the repertoire of the mathematician using c.e. sets and dynamical systems. All one needs are *rules* – not necessarily *axioms*. This is illustrated in the mathematical appendix.

When Simon *et al.* (Levy, 1998b, p. 91) stated the following: 'Chess is the *intellectual* game *par excellence*. ... . If one could devise a successful chess *machine*, one would seem to have penetrated to the core of *human intellectual* endeavour', they seem to have understood that machine intelligence and learnability by machines depended on penetrating 'the core of human intellectual endeavour'. The main thrust of this chapter, and its mathematical appendix is based on humans penetrating the core of the *machine's intellectual endeavour*. Unless research and realizations of machine intelligence and learnability by machines come to terms with the machine's intellectual endeavour, then, like the dependence of algorithmic formulation assuming the Church–Turing thesis, humans will be chasing a will-o'-the-wisp!

Appendix: Mathematical Notes

Theorem

Every acute-angled triangle can be *made into*, or *from*, three isosceles triangles.[28]

Proof

See Figure 2.1.

The above theorem is 'proved' *visually* – without words or symbols. Now, take an example of the *proof* of a Euclidean proposition by a *machine*, given

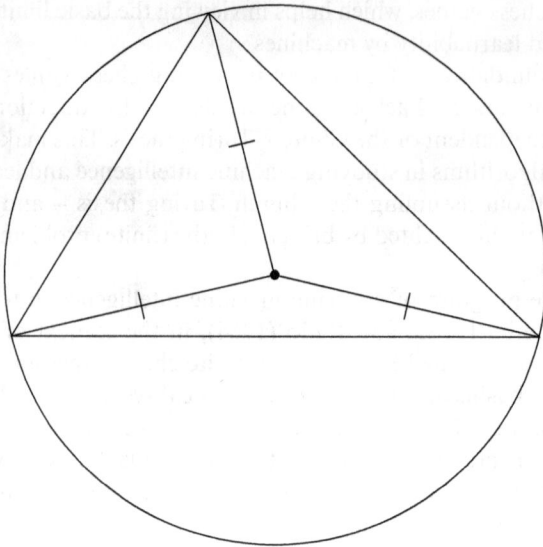

Figure 2.1 Every acute-angled triangle can be *made into*, or *from*, three isosceles
triangles.

Source: Adapted from MacHale (2008).

only the *rules* of Euclidean geometry, using *a program* – i.e., algorithm –
devised by Marvin Minsky.[29] The 'rules of Euclidean geometry' include the
'fact' that the three angles of a triangle must sum to 180°; this is *equivalent to*
the *axiom* of parallels – the removal of which leads to non-Euclidean geom-
etries (of Bolyai, Lobachevsky, Gauss, etc).

Axioms, like the 'character of an opponent', are *human* concepts – like
the axiom of choice, the axiom of completeness, etc. Humans develop axiom
systems, seek formalizations, and the like, to solve problems or generate
theorems. This is, I think, what Turing *meant*, when he stated:

> [I]f a mathematician is confronted with such a problem [i.e., that with
> certain logical systems, there is no test that can be applied which will
> divide propositions, *with certainty*, into classes that were provable and
> unprovable] he would *search* around[30] and find new methods of proof, so
> that he ought eventually to be able to reach a decision about any given
> formula.
>
> (Turing, 1992a [1947], p. 104; italics added)

This is coupled with the idea in Turing (1992b [1947], p. 127; italics added),
that: ' [I]ntellectual activity consists mainly of various kinds of *search*.'

What do *machines* do? Do they *search* over a space autonomously – that is, without human intervention? Do they formulate axioms to facilitate the proving of theorems? Could they have developed the axiom of completeness due to their dissatisfaction with a non-constructive mathematics dependent on the axiom of choice? Can they develop an intuitionistic constructive mathematics by themselves; that is, without the intervention of a Brouwer or a Martin-Löf.

I like to think that this is the defining characteristic between humans and machines – but I am not sure. After all, there are mathematical and physical systems which have defied axiomatization – Bishop's *constructive mathematics*, *Feynman diagrams*, even Brouwer's intuitionistic constructivism (despite Heyting's 'compromises') are examples of the failure of the axiomatizing crusade – which has nothing to do with machine intelligence or learnability by machines.

Author's Note: This is a *Turing–Post* centred piece. In this sense it is a companion piece to Velupillai (2013, 2017a, 2017b, 2020). As far as Simon is concerned, Velupillai (2018) is relevant, especially for The Pioneers –Turing, Shannon and Simon section above but it is my *almost* my first foray into anything by Shannon (except some notes on his work on the *Differential Analyzer* – which is viewed from the point of view of *analogue computing*): see also the interpretation in chapter 4 of Simon, *ibid,* of *his* early work on cybernetics and Velupillai (2004).

Notes

1 Premchand's 1924 short story, 'The Chess Players' ('Shatranj Ke Khilari') was made into a film (of the same name) by Satyajit Ray, in 1977.
2 See Flanagan (2009) on Duchamp and Orozco and Kuenzli & Naumann (1990) for Duchamp, alone; I hope Gabriel Orozco is not 'confused' with José Clemente Orozco!
3 Alessandro De Rosa (2019).
4 There is an *alleged* etching by the young Hitler's art teacher, Emma Löwenstamm, showing Lenin playing chess against the budding Nazi (*The Telegraph*, 3 September, 2009).
5 Boorman (1969).
6 van Dalen (2013), p. 643, where Brouwer is described as being 'a fervent chess player' and this may be the reason for Max Euwe (who defeated Alexander Alekhine for the World Chess Championship in 1935) being his PhD student. Incidentally, many seem to attribute to John von Neumann the 'first' devising, in 1928, of the min-max (or saddle-point) property of a two-person game equilibrium concept; Euwe, working within Brouwer's set theory, developed a constructive min-max solution, also in 1928.
7 The word *deep* is used in the way that moves and strategies of play are evaluated in chess and GO by humans and machines (the latter implementing various optimizing techniques of learning, and searching along, tree strategies, by using neural net structures).

8 There were many others who contributed to the digital computer-based construction of game-playing machines; but I choose to concentrate on these three outstanding scholars, partly due to their competence, but mainly because I am particularly interested in their *oeuvre* with regard to *machine methodology, epistemology* and *philosophy*.

9 Michie refers to the pioneering articles of Shannon (1993 [1950]), Turing (1992d; [1953]) and Newell *et al.* (1958) (see Levy 1988a, 1988b), in the quoted passage.

10 All references to Turing (1992 [a–e]) include, also, the Notes to the respective chapters, of pp. 205–216.

11 I would say that they are also Shannon's views; I am less sure about Simon!

12 I am tempted to think that it is a case of *exclusive OR* (Boole), as against the *inclusive OR* (Jevons); *versus* interpreted as the former and *plus* in terms of the latter.

13 *Machines*, for Turing, Shannon and Simon, did not mean *mechanical*; their machines could be electrical, electronic, even hydraulic and, of course, hybrid (also in being partly digital and partly analogue).

14 Later, from the time of the *Dartmouth conference* (1956), *machine intelligence* came to be referred to as *artificial intelligence* by John McCarthy, Herbert Simon, Marvin Minsky and others.

15 Shannon presented the paper on 'Programming a Computer for Playing Chess' on March 9, 1949 (see footnote on p. 637 of Shannon (1993 [1949]) and Levy (1998a), p.1). However, he *published* it only in 1950 (Levy, 1998a, chapter 1.1; see also Copeland & Prinz, 2017).

16 As Hodges perceptively remarks:

> [W]hat [Turing] needed most was a grip on rigour, on intellectual toughness, on something that was absolutely right. While the Cambridge Tripos – half 'pure' and half 'applied' – kept him in touch with *science*, it was to *pure mathematics* that [Turing] turned...
>
> (1983, p. 61; italics added)

Shannon, on the other hand, as a *scientist*, never harboured any doubts about the rigour of the *engineering approximations* of *his* constructions, nor needed any confirmation on *his* 'intellectual toughness' – Turing sought the comfort of mathematically 'rigorous' approximations of *his scientific* constructions, whether it be the Turing Machine, Enigma machine or the Riemann 'machine', or whatever.

17 See p. 469 of Shannon's *Collected Papers* (Sloane & Wyner, 1993) and Novikov (1964; p. 21 is especially relevant in the Shannon finite construction sense of relay and switching circuits-based chess-playing machines); I am greatly indebted to the insights offered by the latter, especially on the relevance of Hilbert's Finitism to the validity of *tertium non datur* and proof by contradiction. All the pioneers used these techniques freely because they worked, often implicitly, with finite systems for constructing machines. In passing, I would like to mention – especially since the Preface and Notes to Novikov (1964) is written by Goodstein – that his valuable work on *Boolean Algebra* (Goodstein, 1963) was decisively important for me in appreciating Shannon's 1993 [1938] classic.

18 Boole is mentioned on p. 474 of the classic 1938 article on Relay and Switching Circuits (Shannon, 1993 [1938]), but Jevons in none of his articles; likewise, there is no mention of Emil Post (or Wittgenstein) even when he relies so much on the *Truth Table* method. Of course, there is no mention of Hailperin (1981) in 1939.

19 It is impossible to talk of *Herbert Simon*, especially with regard to chess, without also discussing *Allen Newell*! Hence, this section should be read – if at all – *as if* it dealt with both of them (and, in part, also with *Cliff Shaw*, in connection with programming chess machines to play). In addition, I am convinced that Newell is part of the group of classic behavioural economists formed by Day, March, Nelson, Simon and Winter.

20 It is the third (and last) part, of the three issues of cryptarithmetic (puzzle-solving, chapters 5, 6 and 7), logic (theorem-proving, chapters 8, 9 and 10) and chess. Their common framework of analysis is as *Information Processing Systems* (IPS) solving problems as humans would – not as machines do. Hence, the importance of the Logic Theorist as a programming entity and the role of heuristics in a search space of solutions (win, lose, draw) in chess problems.

21 He went on to claim (1970, p. 293): '[W]e do not know any theorem of $T(\mathcal{R})$ which is not game derivable'.

22 A *basic game* is defined on p. 292, Lachlan (1970) involving an r.e. sequence of *requirements* and a *finite* combination of *atomic formulas*. I shall use *computable enumerability* (*c.e.*) instead of r.e. in the sequel.

23 Or chapter 16 of Soare (2016), in which he also states (p. 221): 'By the late 1970s, Lachlan had invented an intuitive game theory model for constructing c.e. sets which clearly revealed the intuition'. Lachlan had, in fact, 'invented an intuitive game theory model' already by the late 1960s, but apart from this, I agree completely with Soare.

24 Recall that *finite* sets are *computable*.

25 See von Neumann & Morgenstern (1953), p. 59, footnote 3 (highly relevant also for the next section) and §15.7.

26 *Undecidable* in a (computable) mathematical or (mathematical) logical sense.

27 Recall that *machine intelligence* is the original phrase for *artificial intelligence*; I prefer the 'old fashioned' phrase, especially because it was *not* introduced to differentiate the concept from *cybernetics*.

28 Adapted from MacHale (2008).

29 Michie (1986, p. 11 and p. 23; the Minsky role is emphasized in McCorduck (2004, p. 126) and also in note 21, p. 352 of MacKenzie, 2001 (except that there is an obvious typographical error in referring to p. 106 instead of p. 126; it may be that MacKenzie is referring to an earlier edition of McCorduck).

30 In his/her mind!

Bibliography

Berlekamp, Elwyn R., John H. Conway & Richard K. Guy (1982), *Winning Ways for Your Mathematical Plays*, Academic Press, London. (Originally published in two volumes: Vol. 1 *Games in General*, Vol. 2 *Games in Particular*.)

Boorman, Scott A. (1969), *The Protracted Game: A Wei-ch'i Interpretation of Maoist Revolutionary Strategy*, Oxford University Press, New York.

Copeland, Jack B. (1999), A Lecture and Two Radio Broadcasts on Machine Intelligence by Alan Turing, edited by Jack B. Copeland, pp. 445–476, chapter 22, in: *Machine Intelligence, Intelligent Agents*, edited by K. Furukawa, D. Michie & S. Muggleton, Oxford University Press, Oxford.

Copeland, Jack & Dani Prinz (2017), Computer Chess – The First Moments, pp. 327–346, chapter 31, in: *The Turing Guide* by Jack Copeland, Jonathan Bowen, Mark Sprevak, Robin Wilson, et al., Oxford University Press, Oxford.

De Rosa, Alessandro (ed.) (2019), *Ennio Morricone: In His Own Words*, Oxford University Press, Oxford.

Euwe, Max (2016 [1928]), Mathematics – Set Theoretic Considerations on the Game of Chess, *New Mathematics and Natural Computation*, Vol. 12. No. 1, pp. 11–20.

Flanagan, Mary (2009), *Critical Play: Radical Game Design*, The MIT Press, Cambridge, MA.

Franklin, Benjamin (2004), *The Autobiography and Other Writings on Politics, Economics, and Virtue*, edited by Alan Houston, Cambridge University Press, Cambridge.

Giannini, Tula & Jonathan P. Bowen (2017), Life in Code and Digits: When Shannon Met Turing, paper presented at the Conference of Electronic Visualisation and the Arts (EVA 2017), held in London on 11–13 July, pp. 51–58. Available at www. researchgate.net/publication/318756069_Life_in_Code_and_Digits_When_Shannon_Met_Turing

Goodstein, R. L. (1963), *Boolean Algebra*, Pergamon Press, London.

Hailperin, Theodore (1981), Boole's Algebra Isn't Boolean Algebra, *Mathematics Magazine*, Vol. 54, No. 4, September, pp. 173–184.

Hodges, Andrew (1983), *Alan Turing: The Enigma*, Burnett Books Limited, London.

Hodges, Andrew (2008), Alan Turing: Logical and Physical, pp. 3–15, chapter 1, in: *New Computational Paradigms: Changing Conceptions of What Is Computable*, edited by S. Barry Cooper, Benedikt Löwe & Andrea Sorbi, Springer Science + Business Media, LLC.

House, R. W. & T. Rado (1964), An Approach to Artificial Intelligence, IEEE Transactions in Communications and Electronics, Vol. 83, No. 70, January, pp. 111–116.

Ioan, James (2009), Claude Elwood Shannon: 30 April 1916–24 February 2001, pp. 258–265, in: *Biographical Memoirs of Fellows of the Royal Society*, Vol. 55, The Royal Society Publishing, London.

Kao, Ying-Fang & K. Vela Velupillai (2013), Behavioural Economics: Classical and Modern, *European Journal of the History of Economic Thought*, Vol. 22, No. 2, April, pp. 236–271.

Kawabata, Yasunari (1972 [1951]), *The Master of GO*, translated from the Japanese [Meijin] by Edward G. Seidensticker, Vintage Books, New York.

Kuenzli, Rudolf E. & Francis M. Naumann (eds) (1990), *Marcel Duchamp: Artist of the Century*, The MIT Press, Cambridge, MA.

Lachlan, A. H. (1970), On Some Games Which Are Relevant to the Theory of Recursively Enumerable Sets, *Annals of Mathematics*, Second Series, Vol. 91, No. 2, March, pp. 291–310.

Levy, David N. L. (ed.) (1988a), *Computer Chess Compendium*, B. T. Batsford Ltd., London.

Levy, David N. L. (ed.) (1988b), *Computer Games I*, Springer-Verlag, New York.

MacHale, Des (2008), Proof without Words: Isosceles Dissections, *Mathematics Magazine*, Vol. 81, No. 5, December, p. 366.

MacKenzie, Donald (2001), *Mechanizing Proof: Computing, Risk, and Trust*, The MIT Press, Cambridge, MA.

McCorduck, Pamela (2004), *Machines Who Think: A Personal Inquiry into the History and Prospects of Artificial Intelligence*, A K Peters, Ltd., Natick, MA.

Michie, Donald (1986), *On Machine Intelligence* (Second Edition), Ellis Horwood Limited, Chichester, Sussex.

Neumann, J. von & Oskar Morgenstern (1953), *Theory of Games and Economic Behavior* (Third Edition), John Wiley & Sons, Inc., New York.

Newborn, Monty (1997), *Kasparov versus Deep Blue: Computer Chess Comes of Age*, Springer-Verlag, New York, NY.

Newell, Allen & Herbert A. Simon (1972), *Human Problem Solving*, Prentice-Hall Inc., Englewood Cliffs, NJ.

Novikov, P. S. (1964), *Elements of Mathematical Logic*, Oliver & Boyd, Edinburgh & London.

Rado, Tibor (1962), On Non-Computable Functions, *Bell System Technical Journal*, Vol. 41, No. 3, May, pp. 877–884.

Sadler, Matthew & Natasha Regan (2019), *Game Changer: AlphaZero's Groundbreaking Chess Strategies and the Promise of AI*, New in Chess, Alkmaar, The Netherlands.

Shannon, Claude E. (1993 [1938]), A Symbolic Analysis of Relay and Switching Circuits, pp. 471–495, in: *Claude Elwood Shannon: Collected Papers*, edited by N. J. A. Sloane & Aaron D. Wyner, IEEE, Inc., New York.

Shannon, Claude E. (1993 [1949]), Programming a Computer for Playing Chess, pp. 637–656, chapter 54, in: *Claude Elwood Shannon: Collected Papers*, edited by N. J. A. Sloane & Aaron D. Wyner, IEEE, Inc., New York.

Shannon, Claude E. (1993 [1950]), A Chess-Playing Machine, pp. 657–666, chapter 55, in: *Claude Elwood Shannon: Collected Papers*, edited by N. J. A. Sloane & Aaron D. Wyner, IEEE, Inc., New York.

Shannon, Claude E. (1993 [1953]), Computers and Automata, chapter 82, pp. 703–710, chapter 82, in: *Claude Elwood Shannon: Collected Papers*, edited by N. J. A. Sloane & Aaron D. Wyner, IEEE, Inc., New York.

Silver, David et al. (2017), Mastering the Game of Go without Human Knowledge, *Nature*, Vol. 550, No. 7676, 19 October, pp. 354–359.

Simon, Herbert A. (1991), *Models of My Life*, Basic Books, New York.

Singh, Satinder (2017), Learning to Play Go from Scratch, *Nature*, Vol. 550, No. 7676, 19 October, pp. 336–337.

Sloane, N. J. A. & Aaron D. Wyner (eds) (1993), *Claude Elwood Shannon: Collected Papers*, IEEE, Inc., New York.

Soare, Robert I. (2016), *Turing Computability: Theory and Applications*, Springer-Verlag, Berlin.

Stewart, Ian (1991), Deciding the Undecidable, *Nature*, Vol. 352, 22 August, pp. 664–665.

Takeuti, Gaisi (2003), *Memoirs of a Proof Theorist: Gödel and other Logicians*, translated (from the original Japanese) by Mariko Yasugi & Nicholas Passell, World Scientific Publishing Co. Pte. Ltd., Singapore.

Turing, A. M. (1992a [1947]), Lecture to the London Mathematical Society on 20 February 1947, pp. 87–105, chapter 3, in: *Mechanical Intelligence: Collected Works of A. M. Turing*, edited by D. C. Ince, North-Holland, Amsterdam.

Turing, A. M. (1992b [1947]), Intelligent Machinery, pp. 107–127, chapter 4, in: *Mechanical Intelligence: Collected Works of A. M. Turing*, edited by D. C. Ince, North-Holland, Amsterdam.

Turing, A. M. (1992c [1950]), Computing Machinery and Intelligence, pp. 133–160, chapter 6, in: *Mechanical Intelligence: Collected Works of A. M. Turing*, edited by D. C. Ince, North-Holland, Amsterdam.

Turing, A. M. (1992d [1953]), Digital Computers Applied to Games, pp. 161–185, chapter 7, in: *Mechanical Intelligence: Collected Works of A. M. Turing*, edited by D. C. Ince, North-Holland, Amsterdam.

Turing, A. M. (1992e; [1954]), Solvable and Unsolvable Problems, pp. 187–203, chapter 8, in: *Mechanical Intelligence: Collected Works of A. M. Turing*, edited by D. C. Ince, North-Holland, Amsterdam.

van Dalen, Dirk (2013), *L. E. J. Brouwer: Topologist, Intuitionist, Philosopher – How Mathematics Is Rooted in Life*, Springer-Verlag, London.

Velupillai, K. Vela (2004), Economic Dynamics and Computation: Resurrecting the Icarus Tradition, *Metoroeconomica*, Vol. 55, Nos. 2 & 3, May/September, pp. 239–264.

Velupillai, K. Vela (2013), Turing's Economics: A Birth Centennial Homage, *Economia Politica*, Vol. XXX, No. 1, April, pp. 13–31.

Velupillai, K. Vela (2017a), Alan Turing's Orthogonal Trajectories: Decoding the PROF, *Interdisciplinary Journal of Economics and Business Law*, Vol. 6, Issue 3, pp. 87–128.

Velupillai, K. Vela (2017b), Algorithmic Economics: Incomputability, Undecidability and Unsolvability in Economics, pp. 105–120, in: *The Incomputable: Journeys beyond the Turing Barrier*, edited by S. Barry Cooper & Mariya I. Soskova, Springer Nature, Cham, Switzerland.

Velupillai, K. Vela (2018), *Models of Simon*, Routledge, London.

Velupillai, K. Vela (2020), Enigma Variations: A Review Article, *New Mathematics & Natural Computation*, Vol. 16, No. 2, pp. 377–396.

3 Artificial Intelligence, Robotics and Intellectual Property

Ruth Taplin

One of the most persistent questions concerning the development of AI is how 'intelligent' the tools of Artificial Intelligence (AI) and Machine Learning (ML) are. We know from the chapters in this book that robots of all varieties created by AI processes can have data input into them, process it through pattern recognition and offer predictions, forecasting and data-based decisions without human intervention. There have also been recent attempts to make chat bots – customer service bots – more human so that they speak offering reassurance when there are problems. 'No coding' is a new development that allows anyone who uses the internet to direct learning machines through written instruction or voice commands to assist with projects such as scheduling meetings or work programmes.

However, can we say with confidence that AI robots are truly independent of human beings in their decision making and autonomous in their actions? In this chapter we will address this issue of AI autonomy in relation to whether AI robots can be considered autonomous holders of Intellectual Property Rights (IPR) and be granted patents. There has been animated debate from an interdisciplinary perspective and through recent international court cases and among scientific experts.

AI Robots as Creators

One of the most noteworthy cases brought as a test case concerning whether a tool of AI can be considered an autonomous creator and inventor of new patents was the case of DABUS (Device for the Autonomous Bootstrapping of Unified Sentience), which challenged whether the legal status of a law written for human inventors can be applied to learning machines.

The cases concerned the attempts by Dr Stephen Thaler who invented the DABUS robot to have DABUS listed as the inventor in two patents he filed in 2018 for a food container and a neural flashing light – Thaler claims that DABUS created these. It is true that AI and ML can, once inputted with data by humans, use pattern detection to produce predictions, models and forecasts on its own after processing the data. Yet it is human input into the learning machine robot that provides the data to produce an outcome. It is

DOI: 10.4324/9780367857561-3

also a human who built and programmed the robot. Therefore, it may be premature to consider a learning machine robot as an autonomous living being that can be granted patents. The robot does not understand what a patent is and could only be programmed by a human to file for a patent.

After looking at all the relevant court cases that were filed by Dr Stephen Thaler to make DABUS the inventor or creator, we shall look at the works of the renowned roboticist Professor Rodney Brooks – mentioned in Chapter One and Chapter Five on RPA – concerning the possibility of a robot ever being or becoming a living system. Then we will assess the views of Professor Takashi Ikegami from the University of Tokyo, Japan, who, as mentioned in Chapter One, created Alter robots, 1, 2 and 3 and seeks to find the elusive 'juice' of life that is currently missing from robots. He argues that when making learning machines, the biggest mistake by computer scientists is that they separate hardware from software which is united in the human brain but not in AI learning machines.

One of the most noteworthy court cases concerning DABUS was when the UK Court of Appeal ruled on 24 September 2021 that Artificial Intelligence cannot be the inventor of new patents. Dr Stephen Thaler, who had created DABUS, had taken his case to the Court of Appeal after losing his case against the UK's Intellectual Property Office (IPO) which refused to award patents to his AI machine. The IPO had requested that Thaler name a person as an inventor which he would not do. Thaler then took the case to the High Court and lost, so he then went to the Court of Appeal. The judgements made by the Appeal Court judges weighed against the idea of a robot being awarded a patent and having the status of an independent inventor/creator by a two-to-one majority.

"Only a person can have rights. A machine cannot", wrote Lady Justice Elisabeth Laing in her judgement, adding, "A patent is a statutory right and it can only be granted to a person". Lord Justice Arnold agreed, writing: "In my judgement it is clear that, upon a systematic interpretation of the 1977 Act, only a person can be an 'inventor'". He noted further:

> It follows that, on the face of the Form 7s he filed, Dr Thaler did not comply with either of the requirements laid down by section 13(2), and the inevitable consequence is that the applications are deemed to be withdrawn. This is not to introduce some new, non-statutory ground for refusing patent applications. On the contrary, it gives effect to the statutory requirements that (i) the inventor must be a person and (ii) an applicant who is not the inventor must be able, at least in principle, to found an entitlement to apply for a patent in respect of the invention. If Dr Thaler were able to establish that the statute did not require the inventor to be a person and that, as a matter of law, he could derive his entitlement to apply for patents in respect of the inventions purely from his ownership of DABUS, then the position would be different. But, given that the applications do not comply with two important statutory requirements,

it would not be right in my view to permit them to proceed, particularly since the statute provides no other mechanism for addressing the non-compliance.[1]

The third judge, Lord Justice Birss, took a different perspective. While he agreed that "machines are not persons", he concluded that the law did not demand that a person be named as the inventor at all. He further wrote, "The fact that no inventor, properly so-called, can be identified simply means that there is no name which the [IPO] has to mention on the patent as the inventor" and the IPO "is not obliged to name anyone (or anything)". Only Lord Justice Birss interpreted the law to leave room for the possibility that no one needs to be identified as the inventor of a patent while not endorsing or mentioning that an AI robot could be named as the inventor.[2]

Global Implications

Dr Thaler has taken his quest to enable an AI robot to be viewed as an autonomous inventor globally.

Ironically, in his country of origin, the United States (US), his claim faced a resounding defeat. The country that produces the most patents in the world agreed with the UK judges who have delivered judgements on some of the most significant patent cases globally. In early September 2021, US District Judge Leonie M. Brinkema upheld that ruling, writing: "As technology evolves, there may come a time when artificial intelligence reaches a level of sophistication such that it might satisfy accepted meanings of inventorship". She continued, "But that time has not yet arrived, and, if it does, it will be up to Congress to decide how, if at all, it wants to expand the scope of patent law".[3]

Both the European Patent Office and the United States Patents and Trademark Office agreed with the UK Intellectual Property Office that DABUS could not be the owner of the patents applied for by Dr Thaler as DABUS was not a person, and in relation to ownership rights, granting DABUS patent rights would throw into disarray all legal definitions of ownership by a living system.

The Australian and South African Intellectual Property Offices both granted Dr Thaler's applications.

However, South Africa operates a depository system which means that only formal basic requirements are appraised by the patent office without any formal review.

The application was referred to the Australian courts, and the judge, on review of the Australian Patent Act 1990, ruled that an AI system or device can be an inventor, but could not be a grantee of or applicant for a patent. This broadened the interpretation of what can constitute an inventor while upholding the fundamental human rights of ownership.[4]

Whether dividing the meaning of inventor status from ownership and applicant has any real impact remains to be seen, as the concept of an inventor that cannot apply for a patent nor own the rights to the invention seems an exercise in futility. It seems it is not only the definition of what constitutes a living system that is subject to debate, but what constitutes ownership rights.

Living System versus Ownership Rights

Within the legal debate concerning the ability for a learning machine to be classified as an inventor and creator is the fundamental question of what constitutes a living system that can actually have ownership rights.

Differing views on this topic are held by living systems expert Professor Takashi Ikegami and Professor Rodney Brooks, robotics entrepreneur, Panasonic Professor of Robotics (Emeritus) at MIT, author of many books on AI and robots as well being an actual inventor of robotic products such as vacuum cleaners . The former argues that life comes before intelligence so that once robots, for example, become intelligent, they will be concerned with survival and protecting themselves rather than destroying human beings. He has created a group of three robots named Alter who can imitate humans and each other, but questions if this is real life. The latter takes a more pragmatic perspective, viewing robots, such as Roomba the vacuum cleaner robot that he created, as assistants for human beings in their old age. Yet Ikegami, who is Professor at the Department of General Systems Sciences at the University of Tokyo, thinks that robots need more complexity – much greater than quantum mechanics – to develop, and ponders the question of what constitutes Brooks' Juice (his term), which is the essence of life.

Professor Ikegami created the three Alter robots which have been on display in Tokyo and at an AI exhibition in the Barbican, London. In my correspondence with him, he notes that he created these robots to explore how learning machines can imitate life when humans interact with them, and to see how robots can imitate each other. He argues that living systems imitate each other through a process of conscious contagion and is exploring how the human mind can be offloaded onto a learning machine. He further postulates that robots are controlled by software independent from hardware and this is a possible reason why robots cannot be considered living systems as the human mind integrates both software and hardware. Professor Ikegami wrote to me:

> Yes, I agree that consciousness is contagious. The whole idea of contagion is humans offloading their mindset to others and it is not through obvious human action or words. If it could be done with robots that would explain a lot about what is 'life'.

He terms this process mindware and postulates that the mind offloading to another being occurs unconsciously without blatant actions or words. The attainment of life has more to do with mimicry and has its roots in contagion which is a chemical process embedded in the integration of brain software and hardware. He notes, "This is [a] process of mindware as I see it".[5]

Robots can produce many benefits for human activity from data processing to forecasting as outlined in every chapter of this book, but they cannot independently solve complex problems, collaborate, design new products and services or imagine new ways of working in the same way a human can. Additionally, Artificial General Intelligence (AGI) is a heavily researched area, but it is in its infancy, with much more development needed before it can become sophisticated enough to do anything more than very simple administrative work quickly and in large volumes, or assisting humans in mundane tasks as described by Rodney Brooks.[6]

Yet the fact that robots are definitely not living systems at this stage in development and cannot create independently of their human creators/programmers does not seem to stop new attempts by owners of machine learning systems from trying to list them as the creators or inventors of products that can be awarded patents.

Implications of Awarding MLs Patents

In South Africa, patent number 2021/03242 was granted by the South African Patent Office on 28 July 2021 to Dr Stephen L. Thaler which named the inventor as "DABUS". The invention was autonomously generated by an artificial intelligence. Yet experts in the field were not totally convinced by this patent award. Doubt centred on the identity of who the actual inventor was. The interrogation process to identify the inventor which is a formal process that includes relevant laws, treaties and practices was not allowed to occur because the applicant utilised these formalities in such a manner as not to allow the South African Patent Office room to investigate identification of the inventor thoroughly.

The second problem highlighted by experts in Intellectual Property Rights, Adams and Adams (South Africa), is that patent protection can only be pursued as a legal right by those persons who have natural and juristic entitlement. AI does not fit into these two categories and cannot be identified as a legal subject. The right to pursue an application for a patent derives from the inventor, but AI cannot apply autonomously as a legal subject and when Dr Thaler applied for the patent, he could not ask the AI machine DABUS for permission to represent the machine legally.

The Australian Federal Court ruling by Justice John Beach on 30 July 2021 (*Thaler* v. *Commissioner of Patents* [2021] FCA 879) was that AI may be named as an inventor when there is an application for an Australian patent by an applicant. In the Australian case, a statement by Adams and Adams noted that the Australian Federal Court

> [H]eld the view that the entitlement provisions of the Australian patent laws, prescribing who is entitled to seek patent protection, can be interpreted sufficiently broadly to include an owner of AI in the case of an AI created invention, without any proof of entitlement being required,

at least on the basis of the ownership of the AI and its source code. In this regard it is worth noting that the relevant provisions include broad language, including that title to the invention may be "derived" from the inventor.[7]

The Australian Federal Court ruling by Justice John Beach stated that "an inventor as recognized under the Act can be an artificial intelligence system or device. But such a non-human inventor can neither be an applicant for a patent nor a grantee of a patent."[8] This once more points to the legal rights problem of an AI machine being able to apply for a patent and for the owner of the machine DABUS, in this case Dr Thaler, being able to ask it in writing for permission to apply on DABUS's behalf. This perspective is in tandem with the European Patent Office's decision that Dr Thaler's patent application for DABUS be rejected because AI machines or machine learning systems to date have no legal rights; this is because they do not have a legal personality unlike a natural or legal person and consequently cannot be granted legal title over their output, in this case the new discoveries being a fractal container and neural flame.

In relation to global patent standards and rulings, the South African decision is a weak one because South Africa is a non-examining country which can mean that patent applications can be filed, but the applications are not necessarily examined to check if the requirements of patentability are met nor are they mandated to disclose prior art. There is also the possibility that the granted patent can at any point be opposed by a third party and with proof be invalidated. This why many examining countries globally and patent experts have found the decision of the South African Patent Office to be an oversight.

India is an examining country but it is unlikely that DABUS would have been granted the legal right to be an inventor. The most up-to-date patent legislation in India is the Indian Patent Act 1970 which does not offer a specific definition of what constitutes an inventor. A patent application in India can be made by a person (or by the government) who is the first and true inventor. If AI was allowed to be listed as an inventor, and there are very few precedent law cases that are related to the definition of what constitutes an inventor, there would be a number of legal rights that would have to be addressed. These can include the transfer of patent rights, obtaining patent rights, opposition to the patent, and the inclusion of AI as a party to the contract. As these rights can only be granted to a person, the patent rules would have to be re-written for an AI system to be included as an inventor or co-inventor. An important point is that of accountability in the case of wrong doing. How could an AI system be held accountable for its own robotic activities? This is again linked to both legal and ownership rights which the DABUS case does not appear to address.[9]

Ownership Rights of IPR

One oft-overlooked aspect of this issue concerning awarding patents to AI machines is the question of the complexity of ownership rights and ethical/ moral considerations.

IPR and physical property are not necessarily equal. Private property is a good or service which gives the owner the broadest powers of ownership without interference from other parties. The owner may be able to transfer ownership through a legally based exchange. In unusual circumstances, usually under a politically authoritarian government, ownership could be forcibly transferred to the state/government. Or as in the case of some countries such as India, the government can be legally defined as a 'person' who can own a patent. How can an AI machine like DABUS be an owner in the true sense of the entitlement, since it would be unable to engage in a legally based exchange because it has no autonomous intelligence to decide what would constitute a beneficial or harmful transfer of its patent rights?

More sophisticated processes such as infringement which would challenge patent rights cannot be understood or deliberated by a machine nor can DABUS represent itself in court.

Furthermore, property rights give the owner very specific powers and benefits such as the rights to dispose of, use, abuse and have a perception of their property, and to deprive an owner of these rights is a violation of the owners' rights. Yet IPR does not mean total control – otherwise immoral and dangerous outcomes can occur for society, so outside restrictions or externalities can intervene to restrict the occurrence of unforeseen events. In the event of such externalities occurring or in the case of infringement, who is accountable? In the event of AI machines being awarded patent rights how can the nuances of such externalities be dealt with?[10]

AI can become very powerful if the actual owners of the machines make decisions that are detrimental to stakeholders and society. How can an AI machine that is granted a patent right make moral and ethical decisions to stop widespread misuse of the patented product? The product itself, such as food containers or neural flames, may not in itself be capable of creating damaging outcomes, but whoever owns the product through IPR is ultimately the person that can be held accountable. Otherwise, it could lead to ridiculous situations in which the person behind the development of a particular AI machine could blame the machine for failures and improper usages with impunity. There is also the much-discussed ethics of attempting to portray AI machines and processes as a neutral, unbiased process. Sundar Pichal, Chief Executive of Google's parent company Alphabet has stated that AI is the most important discovery since electricity. Yet this has not stopped discord within the AI research team named Google Brain. Dr Timnit Gebru, Head of AI Ethics, asked permission to publish a paper showing how hateful and prejudiced language that AI books learn from books and websites could be uncritically absorbed by AI language programmes, including those created by Google. Her request to publish the paper was rejected and she was fired after this incident.

Recently, another researcher, Dr Satrajit Chatterjee, was also fired. He had led a team of scientists in challenging a research paper, published in the scientific journal *Nature* (2021), which stated that computers were able to design certain parts of a computer chip faster and better than human beings. Dr Chatterjee held reservations about this claim and was not allowed to publish a rebuttal paper.[11]

Chinese Advances in AI

According to the Chinese Standards long-term planning government agency, AI is to be prioritised and China has already as many researchers in the field as does the United States and is working to surpass America's number of researchers through this plan.

One of the Chinese developments in AI is one in which Learning Machines are given the role of prosecutors in legal decision making.

The Chinese 'AI Prosecutor' can charge people with crimes and does so, according to the developers, with almost 97% accuracy based on verbal descriptions of the case presented to it.

The AI machine was built and tested by China's busiest prosecution office which is the Shanghai Pudong People's Procuratorate. To date, it can both identify and press charges for the most common crimes found in Shanghai. These include dangerous driving, theft, credit card fraud and causing quarrels. The Machine Learning device has been trained on over 17,000 cases from 2015 to 2020. The *South China Morning Post* reported that the AI programme run on a desktop computer presses charges by finding repeated patterns obtained from human-generated descriptions of specific 'criminal' behaviour. The conviction rate is very high as China's acquittal rate is less than 1%. In a paper published in the Chinese peer-reviewed journal, *Management Review*, the project's lead scientist, Professor Shi Yong, Director of the Chinese Academy of Sciences' big data and knowledge management laboratory, says that the AI Prosecutor allows prosecutors to concentrate on more difficult tasks.

Not everyone is enamoured with the AI Prosecutor as an offence such as causing quarrels is usually used against anyone who has a different opinion and who is considered a dissident.

An anonymous prosecutor from Guangzhou told the *South China Morning Post* that the claims of 97% accuracy are not always the case and there is room for error in the AI judgements. Related to the 'DABUS as inventor' case described above, the prosecutor further noted that if a mistake in judgement occurs, who will take the responsibility for such a mistaken outcome – will it be the designer of the AI machine algorithm or the learning machine itself? In 2016, Chinese prosecutors began to use an AI tool to evaluate the evidence against the suspect and assess their level of danger to the public. In 2017, China's first cyber court was established for digitally related cases, such as those found in e-commerce, to appear through video chat in the presence of AI judges. Yet human judges monitor the hearings and make the final rulings. Professor Shi stated that AI machines do not as yet have a role in filing charges and suggesting punishments. However, he and his team are developing powerful upgrades that will be able to recognise lesser common crimes and file multiple charges against one suspect.[12]

Robots as Living Systems?

In correspondence with Lord Justice Colin Birss, the author of this chapter requested further clarification regarding an AI machine being considered a living system that has ownership rights entitlement. He replied:[13]

(i) What kind of thing can own property? Answer a person. Therefore, since Dabus is not a person it can't own property.
(ii) What kind of thing can invent a patentable invention? Answer (by a majority) a person.

> An interesting comparison is *AP Racing v Alcon* [2013] EWPCC 3 in which the invention was clearly "generated" by a computer. (It went to the CA but not on any relevant issue.) Does Dabus mean that in fact a trick was missed in that case because what ought to have happened is that the defendant should have pleaded that the invention was not invented by a person and so the inventorship statement must have been false and the application should have been deemed withdrawn?[14]

In my estimation, this insightful analysis will drive the AI/ML machines debate for the foreseeable future.

Notes

1 *Thaler* v. *Comptroller General of Patents Trade Marks And Designs* [2021] EWCA Civ 1374 (21 September 2021), para. 148.
2 See the actual judgement for all the relevant details: *Thaler* v. *Comptroller General of Patents Trade Marks And Designs* [2021] EWCA Civ 1374 (21 September 2021) (www.bailii.org). Also see 'AI Cannot Be the Inventor of a Patent, Appeals Court Rules', BBC Online (24 September 2021), www.bbc.com/news/technology-58668534.
3 'AI Cannot Be the Inventor of a Patent, Appeals Court Rules', BBC Online (24 September 2021), www.bbc.com/news/technology-58668534.
4 See www.penningtonslaw.com/news-publications/latest-news/2021/uk-court-dismisses-dabus-an-ai-machine-cannot-be-an-inventor.
5 From correspondence between Ruth Taplin and Takashi Ikegami during the month of February 2022

6 'AGI Has Been Delayed', Rodney Brooks, blog post, *Robots, AI, and Other Stuff* (17 May 2019), https://rodneybrooks.com/agi-has-been-delayed/.

7 'South African Patent Office's Recent Grant of a Patent for an Invention Created by Artificial Intelligence', www.adams.africa/patents/south-african-patent-offices-recent-grant-of-a-patent-for-an-invention-created-by-artificial-intelligence/.

8 'The Latest News on the DABUS Patent Case', Kingsley Egbuono, IP STARS, www.ipstars.com/NewsAndAnalysis/The-latest-news-on-the-DABUS-patent-case/Index/7366, para. 5.

9 www.mondaq.com/india/patent/1122790/south-africa-grants-a-patent-with-an-artificial-intelligence-ai-system-as-the-inventor-world39s-first#:~:text=In%20July%202021%2C%20the%20South,Sentience)%20listed%20as%20an%20inventor.

10 'Thoughts on Intellectual Property Rights: The "Unorthodox" Approach', D. A. F. Vaquero and M. Reategui, *Veritas & Research*, Vol. 2, Issue 2 (2020), pp. 168–179, www.academia.edu/44851164/Thoughts_on_Intellectual_Property_Rights_the_unorthodox_approach; 'Firms' Obligation to Stakeholders: AI and IPR', Gorden Bowen, Richard Bowen and Atul Sethi, *Interdisciplinary Journal of Economics and Business Law*, Vol. 10, Issue 4 (2021), pp. 26–44.

11 'Another Firing among Google's A.I. Brain Trust, and More Discord', Daisuke Wakabayashi and Cade Metz, *The New York Times* (2 May 2022), www.nytimes.com/2022/05/02/technology/google-fires-ai-researchers.html.

12 'Chinese Scientists Create "World First" AI Prosecutor that Will Decide People's Fate', Louise Watt, *The Telegraph* (26 December 2021), www.telegraph.co.uk/world-news/2021/12/26/chinese-scientists-develop-world-first-ai-prosecutor-can-detect/.

13 Taken from correspondence with Lord Justice Colin Birss of 8 April 2022.

14 See *AP Racing* v. *Alcon* [2013] EWPCC 3, www.casemine.com/judgement/uk/5a8ff72d60d03e7f57ea92a2.

4 The Physical Concept of Information and Artificial Intelligence

Victor Bartenev

The fundamental principle of living systems is the creation of information, which means an increase in the information component of the working potential of a living system. Thanks to the creation of information, living systems survive and develop despite the fatal entropic tendency expressed by the second law of thermodynamics. Like living systems, an artificial intelligence system does physical work by consuming energy from the environment and releasing entropic heat back into the environment. However, any AI system does not create new information – it only receives, processes, stores and transmits huge arrays of binary electrical signals (the so-called "big data") in accordance with an algorithm created by a programmer.

The highest stage in the evolution of living systems was the emergence of a super-creative human brain. Specifically, the human brain, based on genetic data and signals received from the senses, constantly creates information, maintaining an adequately encoded reflection of the surrounding world. A sustainable world is in a state of harmony, which means the balance of all information and energy factors, so people experience a sense of harmony inaccessible to AI.

Introduction

With the advent of the first computer-based AI systems, their capabilities were compared to those of human intelligence. However, there is a fundamental difference between the human brain and the computer. The fact is that only living systems (including people and human society) can create new information, while the functioning of AI systems is based on information "from the past" that was previously created by humans and then encoded in computer hardware and software.

Human creativity is nothing more than the ability of people to create new information. Alas, it is still impossible to give an exhaustive answer to the question of what is new information and, accordingly, what is creativity. Such an answer would mean that we have realised the mystery of the origin of life and the functioning of the human brain, but people are far from solving this problem, if such a solution is possible in principle.

DOI: 10.4324/9780367857561-4

Nevertheless, the physical approach to the interpretation of the concept of information presented in this chapter makes it possible to assess the prospects more adequately for using AI in the global economy and in everyday human activities.

The physical concept of information is one of the central components of the economic paradigm described by the author in a book[1] published by Routledge in 2013. This concept was then discussed in more detail in a separate book[2] published by the University of Warsaw.

In accordance with this paradigm, the global economy is a social living system that increases its working potential (total economic value) through the consumption of primary energy from the environment.

Any living system seeks to increase its working potential by consuming primary energy and removing entropy back into the environment. By removing useless entropy, living systems create useful information (sometimes called "negative entropy"), thereby increasing the information component of the system's working potential.

The physical concept of economic value and information is supported by empirical evidence. The updated evidence[3] primarily includes observed trends in the prices of global energies such as electricity, crude oil and grain, as well as the ratio of gross domestic product (GDP) to money supply in the global economy.

Price trends for world energy carriers are qualitatively different due to the two-component nature of the economic value, which includes an objective energy component and a subjective information component.

The Physical Meaning of Information

The physical approach to the interpretation of the notion of information is that information should be considered as an intangible result of the work done by a combined system:

$$\{living\ observer + observed\ system\}$$

In the socio-economic system,[4] society performs the function of an observer, and the intangible result of the system's work is the growth of the information component of the working potential of the economy. This potential, expressed in monetary units, is equal to the total value created (GDP), which includes a subjective information component and an objective energy component. The created information – that is, the increase in the information component of the working potential of the system – is subjectively assessed (in monetary units) by the society acting as a living observer.

An example of the creation of information on the scale of the universe is the increase in the information component of the working potential of humankind because of the work performed by people together with the star system they observe.

Namely, people observe the sun and other stars, which are open thermodynamic systems. Physical work in stars is performed under the influence of nuclear, electromagnetic and gravitational forces, resulting in an increase in the entropy of the universe due to solar radiation. At the same time, people perceive solar radiation not only as a source of primary energy, but also as light signals. By analysing these signals, people do mental work (which is physical work at the neuronal level), accompanied by the entropy extraction, and in this way, they create information. Accordingly, the information component of the working potential of humankind increases along with the increase in the entropy of the universe.

For a living observer, the creation of information is a useful result, not just useless entropy. In a combined socio-economic system, society acts as an observer. The creation of information and the extraction of entropy provide an increase in the information potential of the global economy, which is reflected in the growth of the information component of the world GDP.

If we try to imagine the universe without living systems, then there will be only matter and energy. In such a lifeless universe, there will be no information, simply because only living systems can create it.

Creation of Information in Living Systems

Living systems possess the fundamental property of information creation, which allows them to exist and develop despite the fatal tendency of an increase in the total entropy of the world (according to the second law of thermodynamics). This fundamental property is the "fifth element" of the general principles of living systems in addition to the four principles (openness,

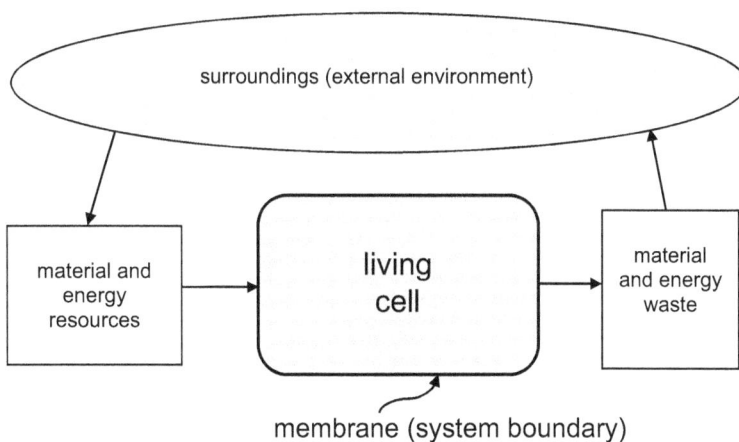

Figure 4.1 A living cell as an open thermodynamic system.

isomorphism, feedback and steady state) formulated by the biologist Ludwig von Bertalanffy in his General Systems Theory.[5]

Information is created and genetically inherited by all living systems. A living cell is the simplest living system (Figure 4.1).

A living cell is an open thermodynamic system, because, first, it contains many molecules in thermal equilibrium. Second, a living cell is separated from the environment by a border in the form of a cell membrane. Third, a living cell receives energy and material resources from the environment, and it releases back energy and material waste.

Multicellular organisms are also open thermodynamic systems; in terms of thermodynamics, the human being is an open thermodynamic system, separated from the environment by the skin.

From a physical point of view, the main result of the work done by a living system is that the primary energy received from the environment is converted into useless energy (entropy) released back into the environment (Figure 4.2).

However, from the point of view of a living observer, a living system is doing work to produce some useful results, rather than producing only useless heat.

The notion of *usefulness* does not exist in inanimate nature; therefore, physics and thermodynamics do not answer the question for what purpose the work is being done. However, living systems usually do work for a specific purpose.

The usefulness of the work done can only be assessed by the living observer who analyses and evaluates the information that is generated because of the work done by the system and the observer (the observation process also requires some work!).

Information and energy are primary notions that cannot be explained in simpler terms. We can only say that energy is an objective value, while

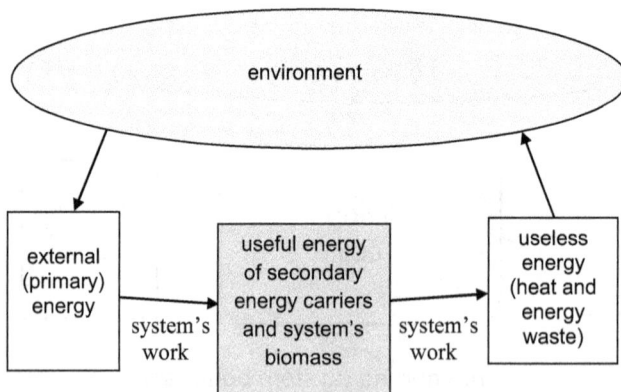

Figure 4.2 Energy transformation in living systems.

information is a subjective value that depends on the living observer. Information exists only in living systems, and life arose with the appearance of useful information instead of useless entropy.

Living systems that do not have a nervous system (for example, unicellular organisms or plants) also perceive, create and transmit information. However, the emergence of the nervous system during biological evolution meant a sharp increase in the information component of the working potential of living systems.

Socio-Economic Evolution

Information growth is a fundamental feature of the development and evolution of living systems. Development and evolution primarily mean improved adaptation to changing environmental conditions.

In accordance with the principle of isomorphism, the most useful properties of living systems are inherited in the process of their evolutionary development, possibly expressed in other structural and functional forms.

Socio-economic evolution is a continuation of biological evolution, and it is rapidly accelerating, resembling a thousand times accelerated biological evolution (Table 4.1).

Thus, the emergence of nerves was a fundamental milestone in biological evolution. Likewise, the creation of electric networks was a crucial stage in socio-economic evolution. Thanks to electrical networks, the economy has been able to create, store and transmit enormous amounts of energy and information at the speed of light. It can be said without exaggeration that electrical networks – power and information networks, wired and wireless networks – are the nerves of the modern economy.

The highest stage of biological evolution was the appearance of humankind. In the same way, the explosive growth of the information potential of humankind, including the development of the Internet and the globalisation of information, can lead to the emergence of a qualitatively new living system as the highest stage of socio-economic evolution.

Table 4.1 The milestones in biological and socio-economic evolution

Biological evolution	Years ago	Socio-economic evolution	Years ago
Emergence of life	~ 3.5 billion	Appearance of humankind	~ 1 million
Appearance of multicellular organisms	~ 1 billion	Emergence of socio-economic relations	~ 10,000
Occurrence of nerves	~ 0.5 billion	Creation of electrical networks	~ 100
Appearance of humankind	~ 1 million	Emergence of Internet and globalisation of information	~ 30

Free Energy

Information is a subjective value that depends on the observer. The information created by a living system does not change the internal energy of the system, but additional information increases the working potential of the system.

In thermodynamics, the working potential (working capacity) is called the *free energy* of the system. The meaning of the term "free energy" is that free energy is equal to the maximum amount of work that the system can do; that is, free energy is the part of the system's energy that is "free to work".

The free energy of an artificial thermodynamic system is expressed in energy units – for example, the working potential of a charged battery is expressed in kilowatt-hours. The working potential of a simple living system performing a single function can also be measured in units of energy. Thus, the working potential of the muscle fibre, expressed in Joules, represents the maximum mechanical work that can be done during muscle contractions.

As for the working potential of a highly developed living organism, it is practically impossible to measure it in units of energy, since it is not clear what should be measured, considering the aspect of "mental work".[6]

The working potential of a social system producing a single product can be estimated as the amount of product produced during the life cycle of the system. Thus, the working potential of a bee colony is the amount of honey produced per year.

The working potential of a socio-economic system producing various goods cannot be measured either in terms of energy or in the amount of any product. Therefore, people created money as a measure of the working potential of the economy.

The thermodynamic system does the work and increases the working potential of the system through external energy consumption; the energy balance is maintained in accordance with the law of conservation of energy. This indisputable fact is called *the first law of thermodynamics*, which can be expressed as follows: *the change in free energy is a consequence of the law of conservation of energy*.

The first law of thermodynamics as applied to the economy means that the economy does work and reproduces its working potential (measured in monetary units as GDP) through the consumption of primary energy. The first law and, accordingly, the law of conservation of energy is monetary, expressed in economic theory in the form of a *balance equation*.

A principal feature of thermodynamic free energy is that it contains two components – useful energy and useless energy. During the working process, accompanied by the effect of friction, the useful energy is transformed into useless energy, and the change in free energy is

$$\Delta(\textit{free energy}) = \Delta(\textit{useful energy}) - \Delta(\textit{useless energy}) \qquad (1)$$

One of the results of the system work is the transformation of external (primary) energy into internal useful energy of the system.

All work processes in a living system are accompanied by friction. Because of this friction, useful energy is irreversibly converted into useless heat, which reduces the working potential of the system. Therefore, useless heat and other waste must be extracted from the system to the environment; the extraction process also requires some work.

A fundamental feature of living systems is that they create information as they work, not just useless heat. In living systems, a decrease in the working potential caused by the conversion of useful energy into useless heat is compensated by an increase in the information component of the working potential. That is, instead of equality (equation 1), we have for living systems:

$$\Delta(\textit{free energy}) = \Delta(\textit{useful energy}) + \Delta(\textit{information component}) \qquad (2)$$

Thus, the creation of information increases the working potential of a living system, and this increase is an intangible compensation for the conversion of useful energy into useless heat.

Note that the free energy components in equation (1) are expressed in energy units. At the same time, the *information component* in equation (2) is immeasurable in energy units because information is not an objective physical quantity. Therefore, the working potential of a living system is immeasurable in terms of energy.

However, in the case of the socio-economic system, people have come up with such a unit in the form of money. Money is used as a measure of the working potential of the economy, so all items in equality (equation 2) in relation to the economy are expressed in monetary units.

Entropy and the Second Law of Thermodynamics

The transformation of useful energy into useless heat during work processes reflects the intrinsic statistical properties of a thermodynamic system. A thermodynamic system is a statistical ensemble that contains a lot of thermally moving molecules. Accordingly, at the micro level, the system has many degrees of freedom. Due to molecular friction, useful energy tends to be randomly distributed (dissipated) over all degrees of freedom – the system tends to the most probable state with a uniform distribution of thermal energy.

The tendency to dissipate useful energy into useless heat is explained by the fact that the return of all molecules back to their original state is statistically unlikely even during the existence of the universe. Therefore, there is a natural tendency to increase the useless energy of an isolated thermodynamic system:

$$\Delta(\textit{useless energy}) \geq 0;$$

this tendency is called the *second law of thermodynamics*.

The second law does not apply to living systems, because living systems are open thermodynamic systems. If you isolate a living system, then it will perish, and only after its death does the second law come into force.

Clausius[7] called useless energy *entropy*. Accordingly, the second law of thermodynamics is often expressed as "the entropy of a closed system increases".

In its current interpretation, the concept of entropy is applicable only to systems that contain a statistically large number of thermally moving microscopic particles, so the state of the system can be characterised by such a macroscopic parameter as temperature. An oversimplified interpretation of entropy as a "measure of disorder" can lead to incorrect conclusions. Thus, the second law of thermodynamics is sometimes inadequately illustrated by the example of increasing disorder in an uncleaned room.

For living systems, the notion of entropy is applicable only at the level of biochemical reactions. However, the notions of working potential, useful and useless energy are applicable to living systems up to the macro level.

Entropy and the second law of thermodynamics are sometimes referred to as the main factors that determine the functioning of an economy. The famous physicist Schrödinger, in his often-cited book "*What Is Life?*"[8] really emphasises the importance of "negative entropy" in life processes. However, Schrödinger notes that a purely entropic interpretation of life processes is an oversimplification for readers inexperienced in physics. Instead of entropy, free energy should be considered.

Entropy and Information

Using the concept of entropy, expression (1) for thermodynamic free energy can be rewritten as follows:

$$free\,energy = useful\,energy - entropy\,component \tag{3}$$

where all equation items are expressed in energy units.

The fundamental property of living systems is that they increase their working potential by creating information, in accordance with expression (2). A living system releases useless energy, and thus the entropy component of free energy decreases. But instead of useless entropy, information is created that is useful for the survival of the system. That is, the working potential (free energy) of a living system can be expressed as follows:

$$working\,potential = useful\,energy + information\,component \tag{4}$$

The principal difference between entropy and information items in expressions (3) and (4) is that entropy is an objective quantity that can be measured by physical methods, while information is a subjective quantity. Information can only be assessed by a living observer; that is, information depends on the observer.

In the socio-economic system, the observer's role is performed by society. Society evaluates the information component in expression (4), thereby assessing the amount of work that needs to be done to manufacture the corresponding products. In other words, society estimates the economic value of goods; such an estimate is expressed in terms of their prices. Accordingly, the price of any product has two components – an objective *useful energy* component and a subjective *information* component.

By doing physical work, a living system creates useful information, not just useless entropy. The creation of information primarily means that a living system, due to the energy of the environment, creates locally (within the boundaries of the system) such conditions in which the second law of thermodynamics is not applicable.

The widespread interpretation of information as "negative entropy" is associated with the definition of entropy as a measure of the uncertainty of the state of a system containing a statistically large number of thermally moving microscopic particles. Entropy increases with increasing uncertainty.

For example, when sugar lumps are dissolved in a glass of tea, the position and velocities of sugar molecules become less certain; therefore, the entropy of a closed thermodynamic system "isolated room + glass of tea + sugar" increases.

From a human point of view, the amount of information about the system decreases as the sugar dissolves. Indeed, before it dissolves, the observer could see that the sugar lump is white and has a cubic shape, but after dissolving, this information "disappeared". Therefore, in human understanding, information is associated with entropy: the less entropy, the more information – therefore information is often called "negative entropy".

In this example, it is essential that the observer is a living person who clearly understands what this means – "white and cubic" – in relation to a lump of sugar.

Information and Maxwell's Demon Paradox

Any observer is a living system that exists in the real world. Accordingly, the observer's size is several orders of magnitude larger than the molecular size. Even the simplest single-celled living system contains a lot of molecules. If we assume the existence of a living observer of molecular size, then such an assumption will lead to thermodynamic paradoxes, such as the *Maxwell's demon* paradox.

The physical concept of information (according to which the creation of information is a fundamental property of living systems) resolves this paradox, which still attracts the attention of researchers.

Maxwell's demon is an imaginary microscopic observer who sits at a tiny hole in the wall that separates adjacent gas volumes 1 and 2. These volumes contain gas molecules that thermally move at different speeds. Initially, both gas volumes are in thermal equilibrium with the same gas temperature.

The observer opens and closes the hole cover, letting in faster molecules from volume 2 to volume 1 and releasing slower molecules back out. As a result, the equiprobable distribution of molecules between the two gas volumes will be violated, and the gas temperature in volume 1 will increase in comparison with the gas temperature in volume 2. This means that the entropy of the closed system {volume 1 + volume 2 + observer} will decrease in contradiction with the second law of thermodynamics.

The impossibility of the above experiment lies in the fact that the receipt by an observer of information about the molecule speed means the creation of information, which can only be done by a living observer. However, living systems of molecular size do not exist in nature. The emergence of life and information began at the level of a living cell, not at the molecular level.

Biological evolution took place over 3.5 billion years, and humankind appeared only 1 million years ago (Table 4.1). Accordingly, genetically encoded information created by living systems during their evolution contains only a fraction of a percent of information that determines exclusively human biology.

In the economy, information exists in different forms and at various levels; these differences reflect evolutionary changes during biological and socio-economic evolution. The information component of the working potential of living systems changes qualitatively during the transition from one evolutionary level to another.

The role of the intangible information component in the working potential of a living system can, to some extent, be illustrated by the example of enzymatic catalysis in biochemical reactions. That is, the information that is encoded by a living cell in the form of the amino acid sequence of the enzyme determines the 3D structure of the enzyme. The enzyme interacts with the initial reagents of the biochemical reaction and orients them spatially so that the energy threshold of the reaction is significantly reduced, due to which the reaction becomes possible.

The enzyme is not consumed during the reaction, and its functioning does not require the expenditure of useful energy. Thus, a living cell creates and encodes information in the form of a DNA nucleotide sequence, which, in turn, encodes the amino acid sequences of proteins, and thus, the system increases its overall working potential.

Subjective Nature of Information

Information is a subjective quantity that depends on the observer. Different observers can evaluate information in different ways; in other words, they can evaluate differently the change in the working potential of the observed system because of the work it has done.

So, at present, from the point of view of human society, the creation of information because of the creation of an industrial infrastructure means an increase in the working potential of the socio-economic system. However, most wild animals negatively assess the importance of this infrastructure for their survival, and they flee away from cities and industries. At the same time, the information potential of dirty urban areas can be highly appreciated by the rat community.

The subjective nature of the information means that the information cannot be characterised only as a "true" or "false" value. This binary evaluation of information is suitable for abstract and artificial computing systems, but not for living systems. For living systems, information can be useful or useless, including all possible intermediate estimates, depending on how the observer estimates the information contribution to the working potential of the combined system

$$\{observer + observed\,system\}.$$

These estimates are subject to change over time. So, Newtonian (classical) mechanics was considered for a long time as the only true theory (that is, as a universally useful theory), but then it became clear that classical mechanics is useless for an adequate description of the microworld, where quantum mechanics is more applicable.

During the life cycle, some of the useless information is removed from the living system in the form of entropic and material waste. However, the remaining useless and even harmful waste accumulates in the living system, thereby limiting its lifespan. As for useful information, its most essential part for the survival of a living system can be genetically inherited. Also, useful information is disseminated and inherited in the learning process.

In the socio-economic system, society plays the role of an observer evaluating the information component of economic value. This subjective assessment is a central factor in the pricing process, which can be adjusted if the demand function is applicable. It is noteworthy that, in principle, the demand function can also be interpreted as a manifestation of the society's assessment of the information component of economic value.

Society evaluates the information component of the working potential of the economy in monetary units. In local economies, these subjective evaluations may differ, which is reflected in the existence of national currencies. Over time, globalisation processes can lead to the disappearance of national currencies and the emergence of a fully adequate world currency.

Information: Units of Measurement

Living systems create intangible information by analysing the received material (physicochemical) signals. This information is encoded and transmitted in the form of other tangible signals.

Material signals can be digitised as binary signals; the total amount of binary signals is measured in bits (or bytes). However, this amount is not a measure of information in terms of the physical concept of information discussed in this Chapter.

Indeed, the computer text of this Chapter contains about 100 kilobytes of binary signals, and the same number of binary signals contains a text of the same size with a random sequence of characters. Of course, the author's work is not at all about typing random characters on the keyboard, which would require an extremely small expenditure of energy. In the case of such a stupid work, the consumed useful energy would only be converted into useless heat, without creating information. This means that the amount of information cannot be measured in bytes.

The author created information because of the work done, which is conceptual (creative) in nature, and thus the author increased his own working potential. To assimilate this information, the reader must also do work and expend the appropriate amount of useful energy in the process of reading. As a result, the information component in the reader's working potential will also increase.

Nowadays, the subjective nature of information is often ignored, and the amount of information is viewed as an objective value that can be calculated in bytes. Thus, the amount of information in the human genome is estimated as the number of possible combinations of nucleotide sequences in human DNA. This estimate is inadequate for the same reasons as in the example above regarding the amount of information in this Chapter.

Biological systems do physical work and create information that is encoded in the nucleotide sequence of DNA. During biological evolution, living systems inherit this information "from the past" through their parents as a basis for creating additional information. If this additional (new) information turns out to be useful – that is, if it increases the working potential of the system and its viability – then this information can be genetically encoded for use by evolutionarily next generations of living systems.

Various biochemical and physiological processes, as well as the instinctive behaviour of living systems, are mainly based on genetically programmed information created by evolutionarily preceding systems.

The highest achievement of biological evolution was the creation of the human brain, which is the most powerful material means for creating information. However, numerical estimates of the information capacity of the brain in bytes are too mechanistic and completely inadequate estimates, suitable for computer networks, and not for the human brain.

There are no objective units for measuring information because the physical mechanisms of its creation and storage in the brain are still unknown, along with the unresolved mystery of the origin of life.

Information and Artificial Intelligence

AI systems are not living systems, so they do not create information. They use information created by people, and this information "from the past" is embodied in the hardware and software of computers. Without human input, this information will become outdated over time.

People make computers that play chess and solve mathematical equations better and faster than people themselves. However, a computer can never write a new physics equation or come up with a new game.

However, children easily come up with new games. During play, children learn to create information in addition to the information that they have genetically inherited or learned from their parents.

The functioning of an AI system is like the instinctive behaviour of living systems. This behaviour is based on generating a pre-programmed response to well-known signals that the system receives from the environment and from other systems. In the event of unknown signals, the AI system becomes helpless.

People need not fear that AI systems will destroy humanity, because AI does not create information. Computer systems are improving rapidly, but at the same rate they are becoming obsolete, and people are throwing them in the trash.

The development of the artificial intelligence and robotics industry is a great achievement in increasing the information component of the working potential of the economy. AI systems and robots do work that does not require the creation of information, thereby helping humans to do more creative work. However, the widespread introduction of robots and AI systems leads to job losses, which is one of the future problems of the world economy.

From Asymmetry to Information

Information arose with the emergence of living systems, and this fundamental event was accompanied by a violation of the principle of symmetry which is observed in inanimate nature.

Indeed, before the appearance of life on Earth, the amount of two mirror isomers of sugar rings and amino acids in the ocean was equal, so the aqueous solution of these molecules did not have optical activity. However, the simplest living cell had to contain only one type of isomers, which are part of the mirror-asymmetric DNA and protein molecules. Life on Earth has chosen to exist in one of the possible mirror worlds.

The entropy of an aqueous solution of organic molecules, measured by the calorimetric method, is the same for a solution containing the selected isomers and for a solution with a random mixture of isomers. However, from the point of view of a living observer, these solutions have different polarisation properties; that is, a living observer can provide useful information about the difference between these solutions.

Some unknown natural processes contributed to the violation of the symmetry of inanimate nature to create living systems that can produce useful information, and not just useless entropy.

The concept of symmetry and asymmetry is directly applicable only to material objects, not to immaterial information. Therefore, the term

54   *Victor Bartenev*

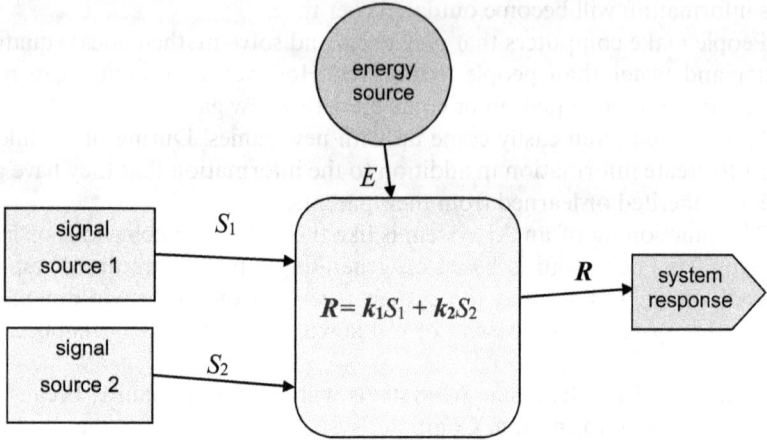

Figure 4.3 Generation of an instinctive system response to signals received from two
well-known signal sources.

"information asymmetry", which has recently appeared in the economic
lexicon, should in fact be interpreted as "inconsistent signals received by the
observer from the same system." Therefore, it is more correct to speak of
"inconsistent signals" instead of "asymmetric information".

People receive incompatible signals, analyse them and create information;
otherwise, they would only act because of instincts based on information
"from the past".

The functioning of AI systems is like the instinctive type of action of a
living system. Instinctive action is the response of a system to certain types
of signals that are well known from the experience of evolutionary preceding
systems.

Namely, let the system receive signals S_1 and S_2, the origin of which is well
known in advance (Figure 4.3).

Then the instinctive reaction is that the control system response (**R**) is
instantly generated:

$$R = k1 S1 + k2 S2$$

where the coefficients k_1 and k_2 have been programmed in advance; these
coefficients are adjusted in the learning process.

The instinctive response requires relatively little energy expenditure.
However, living systems can create new information during their life cycle.
This information increases the working potential of the system, and this
increase requires much more energy than in the case of an instinctive reaction.

If a living system receives unpredictable signals that are incompatible with
previous experience, then the system does work to process these signals to

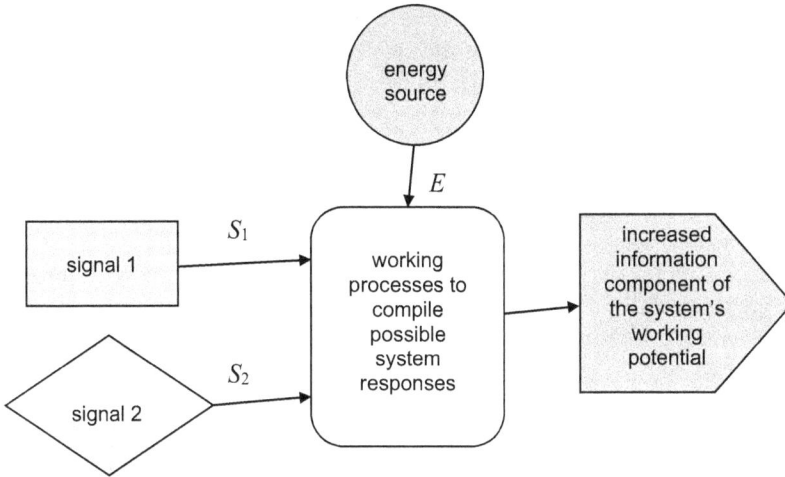

Figure 4.4 Creation of information in a living system based on the processing of two inconsistent (asymmetric) signals.

form an adequate response, which can take a lot of time, even during the next evolutionary generations of the system. As a result, the living system creates useful information that increases the working potential of the system (Figure 4.4).

The creation of information requires much more energy than the energy required for the instinctive response of the system. It is believed that in a few decades, people will be able to create supercomputers with the capabilities of the human brain, which will consume terawatts of energy. However, such supercomputers will simply be extremely powerful signal processing systems with an instinctive type of operation. These supercomputers will help people to create information, but they cannot completely replace the human brain, because supercomputers cannot create new information, no matter how much energy they consume.

Instinctive actions, as shown in Figure 4.3, are actions in an information space familiar to a living system. Creation of information (Figure 4.4) based on the analysis of incompatible (asymmetric) signals means the exit of a living system into another information space with corresponding high energy costs.

As we can see, the emergence of life and the creation of information have something to do with the notion of asymmetry, although we still need to be aware of the root causes of this relationship.

Harmony and AI

The physical concept of information includes the notion of harmony, which is interpreted as the dynamic balance of all tangible and intangible factors.

The global economy, as a highly developed social system, creates information based on the analysis of "asymmetric signals" from all possible sources. The creation of information means the growth of the information component of the working potential of the system. This growth is a principal feature of development and evolution of all living systems.

The development and evolution are not characterised by a desire for static symmetry. Instead, there is a tendency for a dynamic balance, or harmony between all energy and information factors, between all symmetrical and asymmetrical objects.

Static art pictures or material objects (like diamonds) can be symmetrical and beautiful, but they are not harmonious,[9] while dynamic wildlife possesses harmony. Notes of musical scales look symmetrical on paper, but harmony can only be experienced when listening to live music or musical records.

Harmony exists in wildlife mainly due to balanced free energy distribution between numerous degrees of freedom of living systems through dynamic nutritional and metabolic feedback. In the global economy, the distribution of free energy (in monetary units, the world GDP) is far from being balanced, so the current state of the global economy is not a harmonious state.

In fact, there are some signs that the "state of mind" of society is changing in the direction of real feelings for harmony, which is opposed to greed. People understand that harmony is a foremost factor for the unity of man and nature, as evidenced by the enormous success of James Cameron's (2009) film *Avatar*.

The natural human tendency to harmony is particularly evident in people's love of music. Perhaps the electromagnetic oscillations in the human brain correspond to the major-minor rules of musical harmony.[10] If so, then we can assume that music reflects the universal harmony of the world.

Harmonic oscillations underlie universal harmony: they define the stable states of atoms and molecules manifested as the emission spectra; they form the orbits of the planets; they are components of electromagnetic waves in the heart and brain. The global economy is functioning in accordance with the macroeconomic cycle which corresponds to the annual period of the earth's rotation around the sun.

Harmony cannot be numerically evaluated. Nevertheless, harmony is an objective notion. The harmony of the world lies in the fact that, from the micro-level to the scale of the universe, the behaviour of systems meets the objective rules of behaviour of the harmonic oscillator. In the human brain and in human society, these objective rules are possibly reflected in the form of music.

Music has always accompanied the work of people and let them more quickly relax after work. Hard physical and mental work requires musical stimulation: haulers chorus sang their songs, and Einstein played the violin. Nowadays, people can listen to music from their smartphones all day long.

As for AI, it is not able to write a fundamentally new physical equation or compose a memorable musical melody, because the computer does not have a sense of harmony.

People's love for music was expressed in the creation of a theory of musical harmony, which is essentially based on the rules of a harmonic oscillator. It seems that one of the possible natural scientific approaches to fundamental brain research is the Fourier analysis of the brain's electromagnetic waves in comparison with the musical sound waves. Indeed, musical harmony evolved from simple rhythms to classical and electronic music, along with the development of human society; this development first means the growth of the brain information potential.

Unlike the functioning of the human brain, AI operation is not based on the reflection of the harmony of the surrounding world, which includes countless tangible and intangible factors. AI systems can only compile signals and data from a finite number of signal sources strictly in accordance with the action algorithm previously programmed by humans.

AI cannot create new information; that is, it does not have creativity in principle, because the creation of information first of all means maintaining harmony with the outside world and nature.

Creativity versus Algorithmised Behaviour

Any human activity includes not only creativity, but also algorithmic behaviour based on genetically encoded information "from the past" and information obtained in the learning process. As for the AI, it functions strictly in accordance with the algorithm embodied in the computer hardware and software. This algorithm can be refined in an algorithmised learning process.

The ability to create new information or creativity is a fundamental property of living systems, which allows them to evolve in harmony with other living and non-living systems. It is impossible to algorithmise the process of creating new information, because it is impossible to programme the surrounding life. However, there are speculations about AI's ability to create artwork.[11]

Artists paint pictures and musicians compose music based on both algorithmic and creative behaviour. Accordingly, any artwork includes an algorithmised component. Thus, both children and artists usually paint a person with one head, rather than several heads, which reflects their algorithmic behaviour based on information "from the past". Likewise, one of the AI portraits sold at auction for tens of millions of dollars shows a vague image of only one head.

The presence of algorithmised content in music in the form of its rhythm is especially obvious. The genetically encoded rhythmic contractions of the heart predetermined the craving of ancient people for dancing, and then for music.

Both the instinctive and creative behaviours of living systems are evolutionarily improved. The basic principles of instinctive behaviour – encoding and storing information, the formation of an instinctive response to external signals, the learning process – are subject to computer modelling and algorithmizing, and these principles are implemented in AI.

As for the creative behaviour of a living system, it cannot be encoded and thus implemented in AI.

Information appeared along with the emergence of living systems, so that the awareness and algorithmizing of the process of creating new information would first mean solving the mystery of the origin of life. Also, a complete understanding of what new information is impossible without solving the mystery of the functioning of the human brain. Attempts to solve these mysteries of nature using supercomputer modelling are futile, since the process of creating information is not subject to algorithmisation.

The simulation of the origin and evolution of life using a supercomputer and AI is described by a well-known author in a science fiction detective novel.[12] Unfortunately, the author's scientific consultants from famous universities gave him the wrong idea about the possibility of computer modelling of the origin of life.

Namely, in the novel, a supercomputer simulates molecular interactions and molecular evolution, from simple chemical compounds to complex biological molecules. However, in such a computer simulation, an equal number of mirror isomers of organic compounds should appear, but real life is already asymmetric at the molecular level.

Life and information arose along with the mysterious symmetry breaking of inanimate nature. The asymmetry of biological tissues is manifested in their optical activity, first discovered by Louis Pasteur.[13]

Globalisation of Information

According to the physical concept of information, the globalisation of information means the globalisation of the intangible component of the working potential of the world economy. It is obvious that the globalisation of information is far ahead of the integration processes in the global economy; this accelerating lead is associated with the creation of electrical networks (local and global networks) and related information technologies.

The advent of electrical networks and the Internet has allowed the global economy to move into a qualitatively different state, characterised by dramatically increased opportunities for creating information. The emergence of information networks is a fundamental milestone in socio-economic evolution, comparable to the emergence of nerves in biological evolution.

Thanks to the neural network, a highly developed living system can quickly create and transmit information through centrally determined signals that ensure the adequate behaviour of the system. The globalising economy is covered by wired and wireless information networks, and it is developing towards the formation of a unified social system with a common information potential, which is being created by a world community and which is freely available to all members of society.

The globalisation of information will put an end to the pernicious information monopoly of the local establishment and its media, which largely encouraged greed for power and money rather than harmony.

The information potential of the world economy is being globalised in the form of remote education and distance work via the Internet, social networks, online commerce, cloud storage of information, AI systems and other information technologies. Despite all the shortcomings of the current information globalisation, which in an accelerated form repeat the flaws of the previous evolution of living systems, the universal human sense of harmony stabilises this process.

The physical reason for the integration of living systems is that the working potential (free energy) of the combined system is greater than the sum of the working potentials of the individual systems. The globalisation of information is the dominant component of the current integration processes in the world economy.

At first glance, it seems that the globalisation of information leads to a decrease in the role of bright individuals in increasing the information potential of humankind. However, this trend is offset by the formation of a "collective mind" with a dramatically increased ability to create and transmit information using various information technologies, including AI. In turn, new technologies are created primarily thanks to bright individuals!

AI, Big Data and New Information

AI technology is essentially a computer technology for collecting, processing, storing and transmitting data. Recently, instead of the term "data", the term "big data" is increasingly used, which means a set of large amounts of data obtained from qualitatively different sources – from computer databases, text and video files, temperature sensors, online images, etc.

For people and for the world economy, there is no fundamental difference between "data" and "big data", because, first of all, any computer data are just a set of binary signals transmitted and stored in electrical networks. Second, from a programmer's point of view, the transition from "data" to "big data" means only the need to create new software. Finally, it is not about the amount of data (measured in bytes), but about the purpose for which they are used by people.

As discussed above in the relevant section, information cannot be measured in bytes. Accordingly, AI does not create new information while processing big data – it only receives, accumulates, compiles and transmits large amounts of data at a high speed.

Human Brain versus AI

The human brain performs physical work at the neural level by combining big data, which includes genetic data inherited from parents, as well as sensory data obtained through learning and daily activities. As a result of this work, the human brain creates information to maintain an adequately encoded reflection of the surrounding world.

Some *physical principles* of receiving and transmitting signals from the sense organs to the human brain and back to the speech apparatus and organs of movement are successfully implemented in AI systems and robotics. However, the *natural mechanisms* for encoding, decoding, storing and processing these signals in the human brain cannot be adequately modelled based on the physical principles of the functioning of AI and artificial neural networks, since the corresponding natural mechanisms and physical principles differ in accordance with the fundamental differences between living and non-living systems.

The most significant functional properties that distinguish the human brain from AI are the following:

Sense of Harmony

An indispensable condition for the sustainable development of living systems is compliance with the harmony of the surrounding world; that is, maintenance of a dynamic balance of all information and energy factors. The human brain is at the highest stage of the evolution of living systems, and it can adequately reflect the harmony of the surrounding world. AI cannot have a sense of harmony, even though the AI system can be equipped with sensors to obtain all possible data about the environment. Without a sense of harmony, AI is incapable of creative work.

Two-Component Working Potential

The physical concept of information considered in this Chapter complements the author's interpretation of economic value as a monetary measure of the working potential of a socio-economic system.

The working potential of any living system, including the working potential of the human brain, contains energy and information components. The main result of the brain's work is the creation of information; that is, an increase in the information component of the working potential of human society.

The working potential of the AI system contains only the energy component. AI does not create information; all the energy consumed by the AI system during its operation, without human participation, is converted only into entropic heat.

Need for Sleep

To maintain its working potential, the human brain, like the brain of any highly developed living creature, needs sleep. During sleep, the brain continues to consume energy for vital work, part of which is to remove operational information created during wakefulness, which is then useless for the living system.

The concept of the usefulness of information is absent in inanimate nature since information is a subjective value. Accordingly, without human intervention, AI cannot distinguish useful data from useless ones, and it does not need sleep.

During sleep, the human brain can create especially useful information that it could not create during wakefulness. For example, the periodic table of chemical elements and the melody of the Lennon–McCartney song "Yesterday" were invented by their authors in a dream.

As for AI, it is not capable of any kind of creativity, even though it can work without sleep.

Frequency Domain Operation

The harmony of the world, from micro- to macro-level, is based on the objective rules of the harmonic oscillator. It can be assumed that the human brain reflects this harmony by decomposing the space-time audio-visual data from the senses into harmonic components. In other words, the world around us is possibly modelled in the frequency domain of functional Fourier analysis, and not in the space-time domain.

Functional analysis and modelling of system behaviour are much more efficient in the frequency domain. First, a much smaller amount of discrete data about the frequencies of the harmonic components of the incoming signals needs to be stored. Second, mathematical operations are greatly simplified – for example, the operation of integral convolution is reduced to simple multiplication. Finally, with Fourier synthesis of data, previously stored in the frequency domain, back into space-time representation, it is possible to simulate the state of the system at any place in time and space, both in the past (these are memories) and in the future (these are expectations).

Binary signals are processed in logical cells of AI chips in space-time representation. These operations are performed extremely quickly in the nano-range of measurements. Electrical impulses are transmitted much more slowly through neuronal networks. But there are fewer atoms in the universe than the number of possible combinations of interactions between the neurons of the human brain.

If we assume that the electric pulses transmitted during these interactions correspond to different harmonics encoded in the human brain, then such a frequency coding is sufficient to reflect the harmony of the world.

Concluding Remarks

From the physical concept of information, it follows that AI does not have creativity, because, unlike living systems, it does not create information. Nevertheless, AI can contribute to the growth of the information potential of humankind, partially freeing people from non-creative work.

From time to time, interest in the topic of "artificial intelligence" flares up and then subsides again. In the 1970s and 1980s, the term "expert systems" was fashionable instead of AI.

Nowadays, the hype surrounding the alleged breakthrough capabilities of AI is often fuelled with the aim of obtaining funding for related projects without real justification. The term "AI technologies" must necessarily be present in the names of promising projects, just as the term "nanotechnologies" was recently needed. However, the controversial possibilities of AI as one of the areas of information technology should be widely discussed.

In principle, AI does not pose a threat to humanity, because AI does not create information, so it cannot completely replace humans. However, the development and implementation of AI in weapons systems is a deadly threat to humanity. An AI error while driving an unmanned vehicle can lead to the death of several people, while the inadequacy of AI in controlling a nuclear or space weapon is fraught with potential global catastrophe.

The essence of the physical concept of information is that the creation of information means an increase in the subjective component of the working potential of a living system. The human brain, not AI, has always been and will be the natural basis for the working potential of a socio-economic system. In turn, the sharply increased information technology (IT) capabilities contribute to the accelerated globalisation of information; that is, the information potential of the global economy is rapidly forming.

The creation of a common information space will be the highest stage of socio-economic evolution. During the biological evolution, the creation and accumulation of information in the form of a single gene pool of all living organisms was accompanied by the development and complication of universal harmony in nature. As a result, biological evolution did not stop at a more primitive stage – it led to the emergence of people with their sense of harmony.

Therefore, there is reason to believe that the highest stage of socio-economic evolution will be more harmonious than the previous stages. This will be the case if people rely only on their own intelligence, and not on AI.

Notes

1 Bartenev, V. (2013). Value-energy interrelationship and dynamic added value taxation, in *Intellectual Property Valuation and Innovation: Towards Global Harmonisation* (Ed. Ruth Taplin). Oxon and New York: Routledge.

2 Bartenev, V. N. (2018). *Natural Science Foundations of Macroeconomics: The Global Economy as a Living System*. Warsaw: Warsaw University Faculty of Management Press.

3 Bartenev, V. N. (2021). The Physical Concept of Economic Value: Updated Empirical Evidence, *Interdisciplinary Journal of Economics and Business Law*, Vol. 10, Issue 1, 21–43.

4 In this Chapter, the terms "economy" and "socio-economic system" are synonymous.

5 Bertalanffy, L. (1968). *General System Theory: Foundations, Developments, Applications*. New York: Braziller.

6 At the level of functioning of the neurons, "mental work" is a physical work that is done in the nerve cells due to the expenditure of useful energy obtained by the living system from food.

7 Rudolf Clausius (1822–1888) is one of the central founders of thermodynamics.

8 Schrödinger, E. (1946). *What Is Life?* New York: Macmillan.

9 It is said that the face of da Vinci's *Mona Lisa* looks so lively and mysterious because it is slightly asymmetrical.

10 When we forget what we were going to do, we feel a sense of discomfort. A similar feeling arises when we do not hear the completion of a musical phrase.

11 Marcus du Sautoy (2019). *The Creativity Code: How AI Is Learning to Write, Paint and Think*. London: Fourth Estate.

12 Dan Brown (2017). *Origin*. Doubleday.

13 Louis Pasteur (1822–1895) – French chemist and microbiologist who was one of the most important founders of medical microbiology; he pioneered the study of molecular asymmetry in biological tissues.

5 How Robotic Process Automation is Revolutionising Service Industries

Paul Whiteside, Chin-Bun Tse and Amelia Yuen Shan Au-Yeung

Introduction

Digital disruption is changing business models and creating opportunities at an unprecedented rate. Incumbent industry leaders must compete with fast, nimble pure digital businesses with lower operating costs, and achieving business agility based on strong process management foundations is of paramount importance.

Traditional service industries, historically reliant on a human workforce to carry out simple tasks within larger processes, now have opportunities to harness the full talents of their human workforce, while driving process efficiencies, savings, and better quality by using robotics.

In the space of only a few years advances in Artificial Intelligence (AI) and Machine Learning (ML) have been combined with process automation approaches to make advanced robotics accessible to service industries.

Robotic Process Automation (RPA) has been hailed as a game-changing approach that will revolutionise service industries. However, its implementation has proved difficult and, in many cases, RPA has not lived up to its promises, producing undesirable side effects. An estimated 50% of RPA initiatives fail to deliver the expected improvements.[1]

This chapter examines what RPA is, how it's being used in service industries and its contribution to digital business transformation (Digital Transformation). It presents the learnings and best practices that have evolved as the technology goes through its early stages of adoption to become a well-understood part of the digital transformation toolset, with the potential to not only produce efficiencies and reduce costs, but to redefine businesses and propel financial performance.

In a digital age with business agility firmly on the strategic agenda, RPA can help businesses to build better products and services for their customers, and also reshape business models to produce a competitive advantage.

Naturally, this has driven high levels of interest and excitement from the investment community and business executives, but it has also been accompanied by much hype and misinformation as the big market players refine their offerings and jostle for position.

DOI: 10.4324/9780367857561-5

Many businesses have now taken their first steps in RPA adoption and begun to recognise its strengths and weaknesses. In the race to automate, there have been many failures and lessons learned, but also many successes. This is a natural adoption cycle for new technologies, and the lessons for these failures and successes will be discussed later in this chapter.

RPA has the potential to create significant social impacts and there are fears that it could impact traditional jobs by replacing human-centred administrative functions with robotic software agents. In most cases, automation tools only perform low-level tasks, but do so at pace and in volume. These tasks still require human supervision and the associated legal accountability. Human workers will still be needed, but their jobs will evolve to leverage assistive automation which allows more focus on planning, monitoring, process tuning, managing complex exceptions, and, of course, dealing with interpersonal situations where human skills like empathy and problem solving remain massively beyond the capabilities of even the most sophisticated robotics.

RPA will continue to grow and will enable important digital capabilities of future businesses. While RPA is unlikely to take over business functions to the extent that RPA vendors have suggested, as the technology evolves, new use cases and business benefits are emerging.

Defining Robotic Process Automation

RPA is the name of a family of productivity technologies that automate relatively simple and repetitive human tasks. RPA provides a suite of capabilities that enable routine tasks like processing invoices or data entry to be automated with relative ease by non-technical staff.

The robotic techniques at the heart of RPA evolved from screen scraping and testing technologies that have been augmented with basic ML and cognitive abilities. Although sophisticated, RPA is simply not that "smart".[2] RPA changes the way people work, and in many cases makes it more engaging and interesting by automating the monotonous and repetitive supporting activities.

Activities suitable for automation by RPA robots have been nicknamed "swivel chair" processes. These involve manually getting information from a system, then evaluating what to do with it, and entering it into another system. This type of activity when done by humans is slow, error-prone, and stressful for a human worker.

RPA has evolved to include Cognitive Artificial Intelligence (CAI) capabilities and Intelligent Process Automation (IPA). The IEEE[3] defines Intelligent Process Automation as follows:[4]

> *[a] preconfigured software instance that uses business rules and predefined activity choreography to complete the autonomous execution of a combination of processes, activities, transactions, and tasks in one or more*

> *unrelated software systems to deliver a result or service with human excep-*
> *tion management.*

Robots never get sick, don't have personal lives or external commitments, can work 24/7 without getting tired, don't need benefits, and are much faster and less expensive than humans. This makes them an attractive proposition for business leaders who want to solve process efficiency challenges, reduce costs, and increase margins. However, robots cannot do nearly as many things as a human can, and often processes that are initially seen to be ideal targets for automation are found to be more nuanced and complex than originally thought when automation is attempted.

Even sophisticated AI can become confused and fail in circumstances in which a human would succeed on intuition without any need to "think". What is easy for a human can be very difficult for a robot. Robots excel at doing relatively easy tasks quickly and reliably. Humans excel at more difficult tasks, but do them more slowly and with less consistency than a robot.

Robots can produce the benefits of savings as mentioned above, but can't solve complex problems, collaborate, design new products and services, or imagine new ways of working in the same way a human can. Although Artificial General Intelligence (AGI) is a heavily researched area, it's in its infancy with a very long way to go before it becomes sophisticated enough to do anything more than very simple administrative work quickly and in large volumes, as described by Rodney Brooks.[5]

The core capabilities of RPA comprise a simple set of reliable tools based on ML techniques. These include capabilities such as optical character recognition, voice recognition, and natural language processing.

Understanding AI and ML

Before discussing robotics and RPA further, it is worth exploring the differences between AI and ML. These terms are often used interchangeably to describe statistical models that rely on probability and "fuzzy logic" to predict outcomes or perform classifications that in turn power rule-based decision making.

Artificial Intelligence is an umbrella term for the science of creating intelligent machines or software applications using statistical models to mimic intelligent behaviour. It encompasses statistical techniques such as Machine Learning, or Neural Networks that model a simple brain.

ML is a subset of AI comprising a variety of statistical techniques concerned with giving a machine the ability to learn based on data. Machine Learning algorithms automatically learn and train themselves without the need to be programmed or requiring any human intervention beyond initial model training.

ML can be classified as either Supervised or Unsupervised learning. Unsupervised learning trains and optimises an algorithm to perform a task to a predefined measure of accuracy or target output. Unsupervised learning requires no human intervention or training and can be used to find previously unknown patterns and relationships in complex, unfamiliar datasets.

Using ML, machines or software agents can learn from data and make accurate predictions. Examples include recommendation systems, email spam filtering, or sentence completion algorithms that support a human user by assisting the job at hand, making it easier.

RPA solutions comprise a suite of ML algorithms that can work directly with desktop or web-based business tools and can be trained to perform certain repetitive tasks. RPA robots cannot achieve higher levels of intelligence but can detect situations that require a human to resolve them successfully. For example, an automated call center robot can process the tone of voice and vocabulary used by a caller to detect emotions like frustration and fast track the call directly to a skilled human operator.

Virtual Workers

A robot or "Bot" is a software program that mimics human actions and carries out tasks and activities on pre-existing systems or applications. RPA robots or virtual workers are intangible software agents, not people or the physical robots common in manufacturing.

A robotic process is similar to a production line in which individual robots perform dependent tasks in sequence until a unit of value has been produced. In the same way as robotics have been used to great effect in manufacturing industries to drive efficiencies, improve quality, and deliver superior products with lower production costs, RPA is now changing how service industries produce value.

Simple desktop automation utilises a robotic virtual worker that can have a range of simple cognitive abilities, from probabilistic Bayesian decision making to the ability to understand the context of emails and extract key information, read text from images, or estimate the costs of damage by interpreting photographs as part of an accident claim against an insurance policy.

Once a single virtual worker has been trained, the configuration can be saved and used to start identical robot instances that will tackle a large batch of work in parallel. This provides a flexible and fungible virtual workforce that can be deployed to match workload demand patterns. A human workforce, by comparison, would require every individual worker to be trained before moving across departments or onto unfamiliar processes. A human workforce also requires regular skill-refreshing sessions to ensure that processes are understood and followed.

Simple Automation

Simple RPA requires a configuration that addresses every possible processing scenario. Unlike manufacturing, in service industries, tasks and outcomes are more variable in nature with many alternative starting states. A missing field in a data entry form must be handled through a set of business rules. For automation of this type, the robot will rely on its configuration and escalate exceptional cases that have not been predefined.

RPA combines speed, accuracy, and efficiency in handling such tasks and removes much of the need for human involvement. Robots performing a variety of tasks can be orchestrated as a virtual workforce by configuring processes that tie together their inputs, outputs, and dependencies as sequential business processes.

These intelligent components are usually configured into a task flow using a studio or designer application. A user can create scripts by training intelligent agents to undertake unskilled tasks that a human would otherwise need to perform.

Intelligent Automation

Intelligent Robotic Process Automation (IRPA) builds on Simple RPA task automation which merely mimics human work. IRPA adds AI which simulates human intelligence. IRPA uses techniques such as computer vision and Natural Language Processing (NLP) to enable the automation of processes that are more complex and variable and thus cannot be defined easily using a rules-based approach. This broadens the spectrum of tasks that can be automated using RPA, and provides more flexibility.

By using AI techniques, robotic workers can mine unstructured data to find meaningful items and create structured data, with logic based on probabilities rather than binary rules. This allows robotics to be applied to tasks such as classifying incoming customer emails, selecting a suitable predefined response, and making system updates based on the content. Another application is using NLP to derive meaning and context from human language and perform actions as needed.

At present, most RPA vendors are offering IRPA capabilities as a part of a wider suite of intelligent components that can be included and configured into business process workflows. Many vendors work in partnership with specialist partner companies that can implement a variety of advanced AI and Analytics techniques as components to give robots increased cognitive and decision-making abilities.

Enterprise RPA

There are two approaches to RPA implementation: Attended RPA, and Unattended RPA.

Attended RPA deploys software robots that automate interactions with existing desktop applications. In this deployment model, robots work to assist human employees in getting repetitive tasks done very quickly and reliably with only complex or unusual cases being handled directly by the employee.

Unattended RPA is the foundation for Enterprise RPA, and uses robots to automate desktop tasks, but also interfaces behind the scenes with several other systems. Tasks are triggered by specific events such as the arrival of an email, a customer call, or the completion of another process. This processing requires no human involvement, except in dealing with processing exceptions, monitoring throughput, and deploying virtual workers via a console in response to demand.

The automation of simple RPA tasks provides the building blocks of Enterprise RPA. This brings enormous efficiencies to individual workers, but it can be very difficult to scale and enable benefits in a systematic way across a large organisation. Enterprise RPA addresses the needs of large organisations that need to automate large-scale processes; for example, processing huge volumes of complex mortgage approvals at a large lender.

Enterprise RPA differs from Simple RPA in that the robot does not do the work via a desktop PC, but on a server instead, either on-premises or in the cloud. This allows Enterprise RPA to employ a fleet of robots in unison that enables scaling up to huge processing volumes when required. Operational controls, security, audit, and governance can be applied from a single centralised location.

A further advantage of the Enterprise RPA model is that it has a more traditional technology ecosystem, with established software development methods that ensure sustainability, adherence to technology strategy, solution re-use, and service availability that simple task-based RPA does not. This produces similar benefits to well-thought-out and carefully designed manufacturing processes, and brings the benefits of speed and agility at scale.

Impacts and Value of RPA

Technology sales have traditionally been to corporate technology departments; however, due to the accessibility of RPA, and its ease of introduction, RPA is being sold into business teams directly, bypassing the IT function.

Businesses that have in the past felt constrained by slow-moving technology departments and have a desire for a higher degree of business agility and rapid time to value are no longer waiting for corporate IT to deliver traditional systems integrations. The demand for democratised technology is high and business executives are convinced of the rapid benefits that RPA can bring.

Impacts and Concerns about RPA

Intelligent RPA has moved beyond work automation carried out by robotic machines. Machines are tools that increase the productivity of workers. Intelligent RPA has the potential to turn machines into workers.[6] Fears stem from AI and RPA being used together to move up the scale from automation of low-skill tasks to threaten higher-skill tasks that were once seen as safe and secure.

In one example, a radiologist may train for years to be able to identify medical problems from X-rays or other images. AI can now accurately do this. However, a radiologist's entire job consists of more than looking at images, and encompasses many human factors such as an ability to work with patients and as a part of a broader medical team, discussing complex problems, devising plans, and guiding towards a positive outcome. Instead of being threatened, radiographers use this new technology to increase their effectiveness and reduce patient risks.

The rate of AI and RPA adoption is rapid, and it is understandable that people in many occupations may feel threatened. RPA drives the need to create new sources of value and income for humans, and new ways of thinking about business and economics will be required. Robotics can be seen as a risk to livelihood, or as an enabler to help a person get their job done more effectively and create new value for their employers.

Efficiency and Productivity Improvements

One of the most prominent benefits sought by companies adopting RPA is in driving efficiencies in existing operations. The simple logic is undeniable; a robot licence costs a fraction of a human employee's salary and the robot can work 24/7 when needed without slowing down or impacting quality. This is a typical Unattended RPA example where large numbers of robotic processes work against a large but consistent body of work items.

It's easy to focus on these benefits, but the total cost of operation over the full lifecycle of RPA usage must be considered. Often when deploying RPA, a realisation occurs that the tasks being automated are a small part of what the employee does, and automating an entire job role is a much larger and more complex undertaking that challenges the initial cost and benefit assumptions.

Often RPA does not deliver the direct saving envisaged as removing human workers would impact other areas of the business. RPA can contribute to cost savings, but it can also produce new income streams when humans are freed up to focus on value-adding activities. The ways in which RPA can underpin revenue and help win more business are only beginning to be understood.

Quality and Process Governance Improvements

RPA reduces human error and improves the consistency of results. Within finance functions, journal entries and reconciliation activities take many hours of human work to perform and a large amount of time can be lost in identifying and correcting errors. Financial controllership activities have clear steps, defined rules, and are highly repetitive, making them ideal for RPA application.

In addition, the quality of financial governance and controls is increased by using RPA. Every robot worker can be tracked at a very granular level with every record read or updated, and every action taken logged in a database.

A robot cannot deviate from its rule, but can identify scenarios in which a human supervisor is required to give an authorisation. In this way, robots reliably process work in a transparent and fully auditable way with every step recorded. This allows for relevant supervisors and managers to be held accountable for robotic workers, and reduces the scope for malicious or unauthorised data access.

The use of RPA for accurate recording of financial information helps to increase data privacy and security and reduce risks of fraudulent data handling associated with human employees or contract staff.

Wider RPA governance is also required to ensure a balance between digital transformation and compliance risk and control policies. RPA pitfalls and the need for effective governance are discussed later in this chapter.

Technology Democratisation

The adoption of RPA has enabled a democratisation of technology. Non-technical business users can now access and purchase powerful technology platforms and also benefit from active working groups and support networks, reducing the reliance on traditional IT departments and software development projects to enable business change.

The use of RPA requires no programming skills and allows business users to directly apply automation to many aspects of their work. Digital business is very fast-moving and to compete, business leaders are using RPA rather than waiting for their IT department to design, plan, and implement process automation work. As such, the buyers of RPA technology products are predominantly business users rather than IT departments.

Digital transformation is becoming a commonplace strategic initiative for companies. It aims to digitise business functions, enabling them to become more agile and competitive. The ability for a company to reduce costs and redeploy people onto more strategic activities that build capabilities, generate new revenue streams, and create new sources of value has become a deciding factor in business success.

Enabling Business Innovation

RPA can be used beyond achieving efficiency gains. The Automobile Association Ireland (AAI) has rapidly introduced RPA Chatbots that can increase customer satisfaction by creating a convenient and frictionless renewal process that does not require long waiting times to reach a human operator. This has allowed call centre staff to focus their efforts on dealing with more complex quotes quickly, which increases customer retention.

For service industries, processing work at lower costs makes more kinds of work feasible and expands the service offerings. This makes RPA a strategic enabler and, as it becomes better understood, will propel its adopters in innovative new directions.

Enabling Human-Centric Business

Although there are concerns that extensive use of RPA may lead to widespread job losses and negative economic consequences, in many instances, the use of robotics has been welcomed by employees whose jobs may have initially seemed at risk.

RPA removes the need for humans to perform mundane, repetitive, labour-intensive, and detailed work and allows them to focus on what matters, using higher-order intelligence in a much more rewarding way.

People who do this kind of work often experience motivational problems and struggle to engage with their work. This often leads to expensive, disruptive staff turnover and negative impacts on business efficiency.

Since the Industrial Revolution workers have become machines in a wider value creation system. RPA can free workers from unenjoyable work and allow them to contribute in ways that better utilise human skills, judgement, and analytical interpretation. Humans can now engage in more complex activities that align with the company's purpose. For example, freeing up workers from back-office processing to work directly with customers drives the customer engagement and business relevance which are critical to business success in the digital age.

The RPA Market

RPA is growing rapidly and is amongst the fastest-growing software markets. Due to the accessibility of RPA, and its ease of introduction, sales that have traditionally been made to corporate technology departments are now being sold directly into business teams, bypassing the technology function.

Because of this robust growth and ongoing positioning of the main RPA vendors, there has been much confusion about what RPA is. It is often wrongly associated with pure AI, which is a separate area of research and the domain of specialised companies.

RPA does, however, lean heavily on pure AI components to create capabilities that enable robots with cognitive capabilities that can be easily used by businesses.

The various vendors in this market all have distinct strengths, weaknesses, and specialisms. However, there are three distinct leaders in RPA; namely, UiPath, Blue Prism, and Automation Anywhere.

RPA has moved on significantly from its humble beginnings in screen scraping and test automation. It is now an umbrella term for any software that automates business tasks by incorporating a core set of technologies in a packaged and configurable way.

RPA is receiving a lot of attention. Automation Anywhere raised $300 million from SoftBank in 2018, giving it a $2.6 billion valuation. As venture capitalists continue to make substantial investments in the industry, the potential for future growth remains the primary motivator. Many investors see RPA as the fundamental technology that will drive global digital transformation across virtually every industry.[7]

Foundational Robotic Capabilities

RPA solutions rely on robotic capabilities that can be orchestrated using workflow modelling tools. These capabilities or "robotic skills" are the foundations of task automation.

Most RPA vendors supply robots with these capabilities and a wider ecosystem of packaged tools that can be integrated with existing business processes. These ecosystems include additional components to interface into powerful Big Data platforms, Artificial Intelligence, and Process Mining. They also provide networks of partners that can help with digital transformation, education, training, or more specialised requirements such as enhanced cognitive abilities that go beyond the standard capabilities offered.

Computer Vision

Computer vision involves scanning an image to identify and classify the objects in it and using Optical Character Recognition (OCR) to extract text information.

Many businesses still employ paper-based processes. This can happen when, for example, a set of documents has been archived and needs to be added into an online electronic data store. These documents can be scanned into digital images or Portable Document Format (PDF) files and stored, but locating them can be difficult unless some key information (metadata) is extracted and stored in a database.

This would once have been a manual data entry job, in which an operator would be trained to understand the content of various scanned documents, extract key pieces of information, and enter them into a system of record. This work is particularly suited to robots, but only if they can identify documents and turn images of text (and in some cases handwriting) into text that can be stored in a database.

OCR fulfils the task of turning scanned images of letters, legal documents, quotations, etc., into text, but advanced usage includes Intelligent OCR (IOCR) which uses AI to understand contextual information about the document, such as whether it's an invoice or a quotation. More advanced components have been trained to understand even poor handwriting. IOCR adds more flexibility to the robot, enabling it to process a wider variety of cases, identify missing information across multiple-page documents with sections in varying orders, handle variations on layouts, tabular data, etc. Use of IOCR provides an enhanced capability with more accuracy and fewer exceptions and supervisor escalations.

Natural Language Processing

This is the ability of software agents to process unstructured natural language as opposed to processing structured syntax or symbolic instructions.

This capability is widely used to create "Chatbots" and virtual agents that can engage in human conversation. The degree to which this is possible is limited at present, but is developing rapidly, with much academic research finding its way into robotic applications that aim to provide the same efficacy as a human-to-human conversation. The current capability is sufficient to allow a robot to engage with a human on a phone call and quickly undertake the required actions to get a customer's job done, adding significant value in terms of reduced call centre operating costs, shorter call durations, and improved customer convenience, satisfaction, and retention.

Virtual Assistants

Advanced, voice-activated virtual Assistants like the well-known Cortana (Microsoft) and Siri (Apple) help human operators in their day-to-day work by lessening the cognitive load of getting work done.

Humans no longer need to search through several network drives or databases to find records or documents, a process that destroys a human worker's thought process and slows work down. The ability to rapidly access information or reliably complete smaller, distracting work items enables faster completion and higher work quality.

Virtual Assistants do not need to be this advanced, and even simple RPA with limited built-in intelligence can help with collating spreadsheets, finding items in file systems or databases based on relevance scoring, and updating back-end systems. These tasks can be invoked in a variety of ways, from clicking a button to launch a robotic process to work on a large batch of spreadsheets, to being voice activated or automatically triggered when a pattern of events or data is detected.

Virtual Assistants are the most common form of RPA, and the most easily accessible. This model of RPA does not threaten human workers, but hugely improves their performance.

Simple RPA

Simple RPA (as opposed to Intelligent RPA) is a set of techniques used to interact with an existing application via its user interface in the same way that a human user would. This technology has existed for many years and is an evolution of user interface testing techniques known as "web drivers" that can click buttons or links, fill form fields, and read characters or response messages from screens.

By adding studio-based configuration and workflow tools that make it easy to set up task automation by recording manual work (similar to macro recording on an Excel spreadsheet), these testing frameworks evolved into powerful tools for transferring data between existing applications via the user interface as opposed to the "back end" or underlying APIs.[8]

Machine Learning

An intelligent robot can be trained on specific tasks. Data collected from the processing is fed back into the robot with human feedback (supervised training) to create better accuracy. A robot can derive contextual information from unstructured data and can adapt to changes in business processes by being trained rather than needing to be re-programmed as with more traditional forms of automation.

Machine Learning technology allows robots to be trained rather than programmed or configured. Trained robots can recognise and interpret images. Other applications include personalisation or sentiment analysis.

Machine Learning is a foundational set of techniques that can transform the abilities of a simple robot. By leveraging ML components or interfacing to ML algorithms, a robot can gain capabilities to deal directly with humans and understand preferences based on customer classifications or segmentation. Using this information, a virtual Assistant can make recommendations, treat particular user groups in certain ways, or even evaluate weightings that describe the risks of a particular transaction.

A robot can be trained to solve simple business problems such as using alternative data sources upon encountering a missing item on a form. A threshold can be set that determines the amount of autonomy the robot has in problem solving, and when to alert a human supervisor.

RPA Adoption

Technology adoption has become standard in business. Even traditional, non-technology businesses have begun to exploit technology and implement digital business strategies to build strategic competitive advantage.

The high expectation of RPA has put fledgling RPA initiatives at risk of failure instead of laying foundations for success. RPA's ease of adoption and promised speed to savings is compelling, with many businesses investing or planning to invest in RPA adoption. However, RPA does not come without problems.

For many Chief Information Officers (CIOs) who are trying to minimise overall systems' complexity and total cost of system ownership, RPA is far from being a solution that can provide long-term systems' stability. It has significant potential to cause operational brittleness that restricts future business agility. This goes against the core business agility benefits that RPA purports to offer. The ever-worsening fragility caused by rapid application of technology in isolated pockets across a business is known as "technical debt".[9]

Business dissatisfaction with RPA is commonplace, with many initial successes eventually exhibiting undesirable side effects that arise from a lack of strategy and governance. This misaligned and siloed usage results in the inability to scale beyond marginal and localised benefits.

The rate of adoption also varies across industries and corporate functions. The leaders in adoption are currently financial services, where RPA is being deployed in core finance and accounting operations. Banks, insurance companies, telecommunications companies, and utility companies have all shown higher rates of investment, as compared to other industries, in support of digital transformation strategies.

Early adopters that implement RPA at scale tend to be motivated as much by the upside growth potential of AI as they are by cutting costs. AI is not only about process automation, but is also used by companies as part of major product and service innovation. This has been the case for early adopters of digital technologies and suggests that AI-driven innovation will be a new source of productivity and may further expand the growing productivity and income gap between high-performing firms and those left behind."[10–11]

The leading industry for RPA adoption is banking with heavy usage in the back-office and accounting functions. Insurance and healthcare are also adopting RPA heavily.

UiPath has analysed trends in adoption over more than ten years and found that in 2018 more than 50% of companies had planned for significant investment in RPA.[12]

RPA adoption is being taken seriously, and this is only set to continue. Although there has been a great interest in pressing RPA into service, many businesses may be planning investment with an unrealistic set of expectations. This is discussed later in the "RPA Pitfalls" section of this chapter.

If the upward trend in investment continues, it is an indication that businesses are becoming familiar with RPA, getting results, and planning further investments.

Case Studies

The AA Ireland: Renewals Processing with Chatbots

The AA Ireland (AAI) introduced RPA Chatbots, a kind of conversational AI, as a way to manage customer policy renewals, improve conversions on new quotes, and create a convenient way for existing customers to gain 24/7 access to information about their policies.

AA Ireland used AI-powered Chatbots to increase customer conversion, reduce missed webchats by 81%, and cut customer call time from 16 minutes to 10 minutes.[13]

By being available to customers out of hours, AAI got an 11% improvement in conversion rate on generated quotes.

The number of missed live chats decreased as the Chatbots provided an overflow facility for customers when human agents were too busy. This also reduced call agent workload, allowing them to spend less time dealing with routine workflows and more time focusing on and converting the more complex quotes. One such scenario is when human operators have more time to deal with customers who may be considering taking their business elsewhere and a more detailed conversation is required to create a deal that is better for that customer. The Chatbots can identify and fast-track these customers to a human operator, who now has more time to help them.

Complete automation of the entire renewal process for simpler renewals was also achieved. Customers found it very easy to interact with the Chatbot and quickly get what they wanted with minimal hassle. Customers have responded very well to the ease of renewing, and it is a key factor in retaining them.

Administration Automation at AXA UK

In the UK, the insurance company AXA has deployed numerous software robots across its business to help employees process administrative tasks such as claims administration.[14] This has saved an estimated 18,000 hours and £140K in productivity gains.

These Attended RPA implementations help employees to get things done, rather than replacing them completely. The admin bots read customer correspondence and match details to the relevant claims' records. When done by a human, this process could take several minutes, but a robot can achieve it in 30 to 40 seconds.

The RPA robots were carefully introduced and the claims handlers adopted and welcomed the changes as they made things easier for them. Savings have come from helping people to be productive rather than replacing them. The use of RPA provided a 200% return on investment (ROI).

Although very useful in this particular case, AXA has found that the bots are limited in what they can do and cannot replace human claims handlers. Technology concerns about locking in technical debt were addressed, and the company is working towards long-term digital strategies that properly leverage the robotic solution.

Although the robots themselves are quite limited in what they can do, there have been several other aspects of AXA's processes that have been identified as suitable candidates for RPA adoption, with a broader rollout planned for the future. One hindrance is the lack of RPA skills, which does not align to RPA industry claims that this technology can be used by people with no technical skills. In reality, AXA found that domain knowledge and RPA capability building is essential if both short-term execution problems and long-term strategic problems are to be averted.

John Lewis: Cross Function RPA

The department store chain John Lewis has taken initial steps to improve productivity after completing a number of RPA automation initiatives and is now considering a broader variety of value-adding use cases for this technology.[15]

The company has automated some business processes using around 40 robots across different departments including finance, human resources (HR), and supply chain with savings of around £5 million to date.

After initial experiments with RPA across the business, the capability has proved to be a success with many further areas identified to exploit the technology further.

The initial implementation allowed the HR department to improve the way in which employment references were collected. Robots were configured to perform the work of collating information from various sources and updating legacy systems for which there had not been a strong business case for expensive traditional integration. Robotics was introduced to the business in this way, paving the way for executive support and raising the awareness of this technology amongst operational staff.

Building on these early foundations, John Lewis has experimented with higher-value operations such as fraud checks, which can now be done consistently using RPA. As opportunities for robotics continued to be identified, a team was built to ensure that consistency in approach and best practices were being followed. This also provides a route to employing more advanced robotics that leverages cognitive RPA abilities, including customer service agents and Chatbots.

RPA Pitfalls

Since RPA does not require knowledge of the internal workings of other systems and works predominantly via application screens, it's useful when there is a need to streamline clerical operations carried out across a collection of existing corporate systems. For more complex cases involving the orchestration of a larger array of tasks across a complex enterprise, RPA can be a poor long-term choice. RPA is sensitive to changes in business requirements or regulations. When processes fail, they need considerable human intervention both to catch up on the work not done, and, in addition, make the RPA implementation work again.

Misunderstood Operating Models

Often there is a long tail of hidden business requirements that have been learned by skilled operators over many years. These include informal business processes or shortcuts, or undocumented ways of working that only emerge when a system change needs to be re-implemented. Often these can be complex, or in place to get around complexity in other areas, and operators can struggle to articulate why they are needed and are a matter of case-by-case judgement and intuition. The result can be a complex RPA system that takes a long time to automate despite the apparent simplicity of the tasks themselves.

The hype around RPA has led to mis-set expectations that robots can bring new efficiencies to any business process and can replace humans easily and with minimal fuss. Only on attempting an implementation do businesses realise that there are many more factors at play, and that the rapid results are restricted to the simpler cases.

Organisations frequently misunderstand their inherent complexities and seek a silver bullet to create efficiencies. Underestimating the complexity and dependencies surrounding a target process to be automated means that the time taken to eventually integrate and automate a bot can be considerably longer than it was thought at project initiation time.

It's easy for a new automated RPA process to become an isolated legacy system, in its own right, with fragmented workflows and automation processes. Without careful end-to-end planning of the controlling processes, and centralised rules management and monitoring, processes can quickly become elaborate, complex, and understood only by the person who implemented it.

RPA's greatest benefit – its ease of adoption, and ability to quickly get automation implemented – is also its most critical weakness. Applying RPA to problem areas without investigating and optimising the underlying process can itself perpetuate legacy problems and make them even more difficult to resolve in the future.

Expediency to reach a short-term result often causes expensive problems as complexity compounds. Over time, the system's ability to be changed diminishes until the very simplest of additions or maintenance work becomes almost impossible. This is the exact opposite of what RPA set out to achieve, and much care and planning must be undertaken to implement RPA sustainably.

Ignoring Root Problems

RPA allows rapid deployment and solves business problems at the point where they have manifested themselves. Using RPA in this way can give rise to rapid but superficial "sticking plaster" solutions. Whilst these solutions can reduce costs and improve efficiencies for simple cases in the short term, they can become very costly to maintain and support in the future.

Often the underlying business dysfunction is ignored or remains undiscovered. The resulting technical debt is a compounding problem that over the years can lead to the need for an expensive re-implementation. Great care and planning are needed to avoid technical debt, and this goes against the value proposition of RPA.

The misapplication of RPA leads to long-term and expensive problems. Although a sense of progress can be encouraging at first, the business is constraining itself to legacy applications. RPA drives other software applications using their user interfaces by reading and writing screens, which means that future upgrades to these applications become very difficult.

Automation Sprawl

In the absence of a structured plan for adopting RPA, larger businesses can fall victim to "automation sprawl" whereby different parts of a business select different tools with overlapping capabilities and approaches to automation.[16] Automation tools aim to create reliable and consistent workflows, and often focus on creating platforms and products that encapsulate best practices within a particular domain. RPA is a generalised tool that allows users a high degree of freedom in implementing automation solutions.

It's easy for a user to select the wrong type of automation technology and deviate from best practices, resulting in an overly complex solution when careful selection would have provided the solution in a more consistent and less bespoke way. This can be wasteful in terms of licences and maintenance and causes confusion across groups.

Different business units or departments may use RPA in an ad hoc way that does not drive synergies across the entire business. This leads to localised pockets of RPA maturity, and some parts of the business struggling with problems that have already been solved in other parts of the business.

Governance and Compliance

Although RPA can be introduced quickly, without an appropriate level of oversight, several risks emerge. These can all be mitigated, and approaches for doing so are discussed later in this chapter.

Security is a concern for many RPA implementations. Often robots are set up as users, with usernames and passwords, and often have access to sensitive information. A robot is just a piece of software and can't have any malicious intent of its own, but the way that passwords are assigned and access rights granted should comply with wider security and governance policies.

The ways that robots are deployed across the business must also be given attention. As a robot is just a software component, it's wise to ensure that strong Service Delivery Management (SDM) is in place.[17] Establishing an RPA Centre of Excellence to help with these concerns is a good idea.

Expectation Management

Some vendors have set unrealistic expectations on ease of deployment and operational effectiveness. Wholesale transformation and digitisation of a business is a large and complex undertaking – and not necessarily just to implement technology. Large-scale business and impact analysis must be done, and effective change management is essential. Some processes are much more complex than they appear on the surface, and changes have impacts across many parts of the business.

Early adoption of RPA should trial small, self-contained, targeted pilot programmes with a rigorous post-implementation review of expected benefits. Frameworks for managing the journey to value with RPA, using a strategic approach and maturity frameworks, are described later in this chapter.

Considerations for Successful RPA Implementation

Approached in a structured way, RPA can provide a large return on investment, yet according to EY, a high proportion of RPA initiatives fail.[18]

RPA is an accessible technology that is targeted at business users rather than technology departments. Despite the ease of initial adoption, becoming a mature RPA business is challenging, with myriad wider implications that require careful consideration. Initial hype has given way to a more realistic understanding of the long-term pros and cons, and the best use cases for creating value through RPA.

Many businesses are inherently large and complex; therefore, as the adoption of RPA grows and departmental processes begin to overlap, the simplicity of automating individual tasks gives way to the complexity of enterprise operating models. Achieving maturity with RPA requires strong business analysis and process engineering skills once the initial early task automation has been done.

Digital Strategy Alignment

Establishing the correct RPA strategy requires careful planning. Considerations include whether RPA will serve specific business automation challenges within departments, or attempt to leverage cognitive capabilities in processes that cut across several business domains and aim to transform the entire business.

RPA has huge strategic potential, beyond its early adoption cases that focus on finding efficiency "quick wins" and reducing costs. Business innovation can be made possible through a strategy that addresses sourcing, stakeholder management, governance, change management, and capability development.[19]

Building an RPA capability with staff trained to understand how to use the technologies, set up and train robots, and conduct business process re-engineering will be necessary. At higher levels of RPA maturity, a coherent approach across many parts of the business will dictate the chances of success.

Cost reduction is a primary driver for investing in RPA; however, RPA maturity can also add business value and create capabilities that ensure business agility. The RPA strategy should aim for a focus on long-term value creation, as premature or ad hoc initiatives can accumulate into an expensive portfolio of RPA implementations.

Implement a Robotic Operating Model

An organisation adopting RPA will need guidelines, a centre of excellence, and an understanding of patterns that achieve consistent results. The Robotic Operating Model (ROM) provides such a framework and ensures that RPA initiatives align to the broader needs of the business. Capgemini have described a set of best practices for the successful deployment of robotics programmes.[20]

As the complexity of an RPA implementation grows, the same problems that come up in classic technology systems start to manifest themselves. This happens when the organisation changes faster than RPA implementation teams can keep pace with. In the ever-present push for expediency and rapid time to value, the long-term total cost of ownership is ignored, impacting business agility as the costs of change increase.

Stability can suffer and changes made to the robotic implementation risk breaking existing logic that the business relies on. For this reason, technology governance and a change roadmap are extremely important, as is bringing technology personnel into the decision making as they will have the key skills of evaluating, monitoring, and maintenance.

Involve In-House Technology teams and Consultants

The implications of RPA can be wider than the initial implementation of task automation. It is easy to create a brittle and unreliable process automation if the implementers do not have some systems knowledge. This may not necessarily be software development, but process skills will be required for anything more than simple task automation. The maintenance and total costs of the ownership exercise can also be neglected in the rush to implement RPA. Technology departments usually have core expertise in evaluating investment cases, vendor selection, security concerns, process design, and the total costs of ownership over many years of operation. The Information Technology (IT) department can help to prevent mistakes from happening if they are brought in at an early stage. Even if the IT department personnel will not be responsible for running the implementation projects, they can provide invaluable contributions.

Without alignment to overall enterprise architecture, or proper consideration of the business's operating model, RPA implementations become an impediment.

Many productivity systems allow users to easily adjust workflows and add or remove widget components from the applications. This will break the robotic processing when, for example, it can't find a button that has been moved or the position or content of a drop-down list changes. If a company uses Software as a Service (SaaS) applications such as Salesforce.com, updates of this kind that happen without announcement could confuse an RPA robot, but be easily navigated by a human operator.

Establish Strong Change Management

It must be ensured that senior leadership support RPA and are aligned on the benefits to individual functions as well as those to the wider operating model. Expectations must be managed and tempered with careful evaluation and testing of early implementations to ensure that the business cases deliver the expected value within budget. There is a long way to go before robots can perform some of the more complex and intuitive business functions currently undertaken by humans that require careful judgement and a deeper understanding of the business. Getting buy-in can be difficult in non-trivial cases that take longer to realise benefits. Building a long-term robotics capability can be a hugely beneficial part of a strategy, but in the early days, initial poor results can hinder adoption and executive sponsorship

Winning support from the workers themselves, who may feel very threatened by the introduction of robots, is required. Neglecting to manage change properly can lead to difficulty in managing morale problems and resistance.

Analyse and Re-Engineer Existing Processes First

Technology and Business Process Re-Engineering professionals have for many years known about the dangers of automating a bad process. It's important to seek to understand how existing processes contribute to your overall operating model and identify any systemic inefficiencies first before applying robotics. Process discovery techniques, such as process mining, can derive insights from existing processes to pinpoint bottlenecks and inefficiencies.[21]

This approach can also contribute to a strong business case that quantifies problems, costs, and benefits. By also defining the desired benefits of process automation in terms of measurable Objectives and Key Results (OKRs),[22] the overall success of an RPA initiative can be tracked and adjusted if needed.

Establish a Centre of Excellence

Building out a robotics capability means that a core pool of expertise is needed. Although this team doesn't need people with strong programming skills, the defining characteristics of a CoE are in its systems engineering and analysis capabilities.

An RPA Centre of Excellence (RPA CoE) will ensure that there is a strategic fit for RPA initiatives, establish best practices and consistency in approach, help different parts of the business to come together, and rapidly find the correct RPA solutions. The role of the RPA CoE is examined in detail in a paper by Sorin Anagnoste.[23]

Achieving RPA Maturity

An ad hoc approach to implementing RPA over time can lead to a fragmented and difficult-to-maintain set of systems that may require more licences than at first thought. By beginning in an exploratory mode to learn the basic building blocks of RPA, and then moving through stages of maturity in a controlled way, these risks can be mitigated.

Although an industry-agreed maturity model for RPA does not exist at present, Capability Maturity Model Integration (CMMI) is a proven framework for achieving process excellence over time. Using CMMI version 1.3 as a process management and behavioural model can provide insights for baselining and optimising organisational capabilities, streamlining and encouraging process improvement, and reducing risks in product or service development through robotics.[24]

RPA maturity depends on the needs of the business and the degree to which they wish to exploit it. Companies can still get significant benefits at the early stages of adoption.

The leading RPA players have seen the difficulties that arise in moving quickly from initial experiments to using RPA at scale, and have devised models and approaches to help guide the process of adoption, commonly referred to as RPA Maturity Models. Although several maturity models exist, they describe similar pathways from initial adoption to transformative enabling capabilities.

A generalised description of RPA adoption starts with initial experimentation that provides clear, easily achieved, and valuable results. Over time, the desired benefits become more ambitious and require high levels of support and planning.

- *Early Adoption:* Early adoption may utilise virtual Assistants extensively, with robotic workers working alongside individual human workers. Implementation is easy, and the benefits often come quickly.
- *Processing at Scale:* Following initial successes, businesses may consider multiplying the workloads that robots can perform and explore Unattended RPA for processing volumes of work. At this stage, more analysis is required and implementation becomes more involved. This kind of initiative is often undertaken within a specific business function or department.
- *Cross Enterprise Capability:* As the understanding of RPA's capabilities and how it can be applied in certain functions grows, businesses will begin to consider how to move the RPA approach to work across more business functions, or within core functions at scale as an RPA foundation used by the entire enterprise.
- *Industry-Level Change:* Success in making RPA work across the business leads to considering how RPA can change the business more

fundamentally, and how RPA-enabled businesses can innovate to develop a new strategic advantage within their industry.

Initial results can be easily achieved, but to maintain reliable delivery of benefits at scale requires strategic planning. The importance of a strategic approach to RPA has been presented in detail by Willcocks, Hindle and Lacity who describe the following phases of adoption and maturity.[25]

The Initialisation level has a focus on experimentation and learning whilst deploying RPA within existing processes and delivery methods. The implementations are not deep and transformative, but exist at the edge (or surface) where operators currently perform operational work.

At the Industrialisation level, the value of RPA is understood, and further commitments and investments are being made in the technology.

The Institutionalisation level is characterised by strategic commitments to use RPA as a transformational technology within a digital strategy, with a focus on creating a fundamental change in the commercial and delivery operations of the business.

RPA in Digital Transformation

A digital business exploits technology as a source of advantage in both its external products and services and its internal operations. Businesses that have not made investments in technology are finding it ever more difficult to compete in a fast-moving digital age in which customer relevance and engagement are key contributors to success. Companies with efficient operations and a strong focus on making things relevant, easy, and convenient for their customers are beating their competition and establishing a lead that becomes ever more difficult to close.

Digital transformation has become a very important strategic initiative, but it can be very difficult to re-engineer legacy systems and undergo business process re-design at the speed and costs required. RPA can allow more rapid implementation of business change with less risk. Its use is not without some drawbacks, however, as the speed of implementation can result in a system that further increases the reliance on inefficient legacy apps which will eventually and inevitably need replacing.

RPA helps to create efficiencies in repetitive tasks. On its own it can't be the single technology that empowers the digital transformation of a company. A high degree of consideration is required to determine the benefits being sought by a digital transformation. RPA is a powerful tool, but it needs to be augmented with other technologies, and implemented as a part of a broader and more strategic automation approach.

Digital technology is increasing competition and disrupting industries. Businesses are being forced to re-evaluate and clarify their strategies, develop new operational capabilities, and deploy their staff onto higher-value activities.

True disruption is not feasible for many incumbent companies, but there is still a need to digitise their businesses and find innovative ways to win and retain customers. This requires the formulation of digital strategies and committing to large-scale change that will enable sustained competitiveness. A critical factor is identifying and investing in relevant technology capabilities that equip the organisation to sustain improvements over time.

Companies need to align their organisational structure to new ways of working. Whilst RPA is a useful tool, it can only aid digital transformation, not replace the need for it, and it is certainly not a shortcut to a high-performance digital company. Without the right strategy and understanding of the business goals to be achieved using technology, there is a danger of putting too much faith in one technology and not adopting the coordinated changes required across the business.

A lack of alignment about where the business needs to go and what role technology will play often results in piecemeal digital initiatives. One prominent risk of RPA is that its initial ease of implementation gives rise to disjointed digital transformation initiatives across the business.[26]

However, RPA is a powerful way to enable digital transformation if planned strategically with the requisite analysis, re-engineering, and architectural oversight that will ensure it becomes a flexible and adaptive part of the organisation's operating model without unnecessary side effects or long-term impediments to business agility. Identifying the root causes of a company's efficiency problems and using RPA to digitise these can be truly transformational.

Often this involves looking across the entire organisation for process improvement opportunities rather than restricting efforts to particular domains that are more forward-looking and willing to invest.

Breaking out of traditional functional silos is a fundamental aspect of digital transformation. By restructuring teams and processes around key strategic initiatives that introduce efficiencies and build value, the core digital capabilities that characterise a successful digital business are established.

These capabilities involve creating new digital experiences for customers at the boundary of the organisation, and digitising the internal operations to create efficient, agile, digital ecosystems on the inside of the organisation.[27]

The Future of RPA

Robotics is far from new, but intelligent RPA applied to service industries has become very common in the last five to ten years. Robotics in its most recognisable form occurs in manufacturing industries where robots excel at activities that need to be performed to a high level of precision, involve moving heavy objects, or require high and rapid volumes of throughput. In these industries humans now focus on product design, thinking of ways to make the processes and quality better, selling the products, and providing enhanced levels of service to customers.

Similarly, service industries have an opportunity to deploy a valuable but expensive human workforce on activities that warrant the costs. In this way, service industries can focus their attention on driving new value and delivering better products and services to customers instead of inwardly looking at operational efficiencies alone.

It has taken some time, however, for businesses to see beyond the immediate quick wins of faster and less expensive operations.

> *The ugly truth surrounding the first seven years of RPA adoption is that we've simply succeeded in using RPA to move data around enterprises faster with less manual intervention rather than to rewire our business processes and create new thresholds of value.* [28]

RPA will certainly change things and replace many existing jobs. What happens next? A new thinking and mindset are needed as humans adjust to competition from robots. Instead, robotics should be embraced and exploited by humans. If humans had better tools and less mundane work to do, what could be achieved? There will be many new jobs created, just as the last 20 years have seen new jobs come into being because of new technologies, desktop PCs, the internet, mobile technologies, social media, etc. It must be remembered that new technology is designed to take on the heavy lifting and liberate humans from mundane processing to more engaging, motivating work. The threat of automation should be seen as an opportunity for human augmentation. [29]

RPA has matured, having gone through an initial period of early adoption excitement and hype to become a better-understood set of technologies. Many businesses that have begun their journey with RPA have discovered a valuable tool, but not a silver bullet. Technology-led business transformation is not as simple as buying RPA tools. Although value can be found quickly in isolated pockets of businesses, longer-term sustainable and fundamental value is more difficult to achieve, and as discussed previously, requires rigorous planning and strategic thinking.

RPA has passed its inflection point and moved beyond the initial hype that sold it as a self-contained, universal fix that needed no maintenance and could replace expensive humans. Like any technology, it has huge benefits, but making it work at scale has proved challenging. With the correct approach, however, scale can be achieved, and many businesses are successfully implementing RPA at scale, having learned lessons and sought or built the right support. RPA is now being used more effectively in rational and well-understood ways.

Summary and Key Observations

The World Economic Forum (WEF) describes a Fourth Industrial Revolution that brings about fundamental changes in the way we work, use technology in our lives, and relate to one another. This new chapter in human development has been enabled by the rapid adoption of far-reaching technology innovations. The speed and scope of this adoption force businesses to reconsider how value is created and present opportunities for our society to become more human-centred.[30]

RPA is moving beyond hype, and some of the fears it generated. Many businesses are harnessing RPA in human-assistive ways. In many cases, RPA automates discrete tasks that are distracting for humans but can also be orchestrated within a wider process to create value. It's extremely difficult to automate every aspect of a human worker's job, and humans will be needed to perform further automation, train robots, improve automated processes, and supervise robotic processes. More than ever the need to identify and fix the root causes of efficiencies, as opposed to applying superficial cost savings, is being understood.

There are some myths about RPA, which upon closer inspection, reveal alternative and interesting opportunities.

Myth: RPA is Mainly Driven by Cost Savings

Although cost savings can be a key driver for RPA implementations, in many cases, after the initial adoption, the technology is seen in a more strategic way that can enable longer-term operational benefits and helps to build innovative new business models. Despite the hype, RPA is proving to be an important extension to the human workforce that does what robots should; repetitive jobs that hold humans back.

Myth: Robots Will Take People's Jobs

The majority of companies that have adopted RPA have introduced new efficiencies to an existing human workforce rather than completely replacing it. The term "robot" was the catalyst to driving huge interest in RPA since its inception. However, most RPA engagements today are still Attended desktop processes that comprise barely more than five robots, as opposed to the Unattended engagements that were the true initial intention when the solution was invented.

Myth: RPA and Full Automation Produce Zero Errors

Although robots can perform well-defined jobs with 100% accuracy, unexpected events or inputs will cause them to fail. Developments in cognitive RPA can help to improve the accuracy, but there is a long way to go before we have 100% automation and robots that can perform like a human.

Myth: Robots Can "Think"

In truth, developing AI to the extent where machines can reason will take a great many years. The intelligence implied is very low-level and restricted to understanding unstructured text or speech, classifying documents, and applying "fuzzy logic" in place of traditional scripts that need to be prescriptive to cover all eventualities.

Myth: RPA is Easy

RPA has not delivered on many of its promises, and the route to success in RPA requires many considerations as outlined previously. RPA is still in its early stages of becoming autonomous and, as such, a change in one part of the process causes a robot to fail in another area. Poorly executed implementation of RPA solves a problem in one area and creates bigger problems in other areas and undesirable long-term side effects that compound the problems RPA was supposed to solve. The need for strategy, ROM, Maturity Frameworks, and an overall better understanding of the cross-discipline collaboration required to make a success of RPA is now much better understood.

RPA, although powerful, is not advanced enough to replace humans and automate entire departments or business functions. Automation is not new, and businesses have been using digital and mechanical technology for decades. This has led to an explosion of economic growth that has created new jobs and roles that did not exist ten years ago. It's exciting that robots are helping to reduce drudgery and help people to focus on the things that matter most to the business.

Early adopters have come through initial implementations with a better understanding of what RPA means in their industry and how to derive an advantage from it. The mindset has moved from a focus on cutting costs to looking across the business for the right targets for RPA implementations that will cut costs, increase productivity, and empower employees to focus on high-value work that propels the business forward and provides a true advantage.

The RPA industry is maturing beyond its early stages and is showing strengths in specific areas. Many organisations are embracing RPA and learning how to use it well, having moved beyond the hype and into a progressive phase of effective adoption.

Just as in manufacturing – RPA in service industries has produced efficiencies and quality. Over the last 50 years, traditional robotics have changed manufacturing beyond recognition, raising the bar on quality levels, and making products accessible. It has allowed people to own better, but less expensive, cars, home equipment, and computers that would not have been possible using manual building techniques.

RPA for service industries, when deployed in a strategic way and as part of a digital transformation agenda rather than for quick savings, has now reached a level of maturity where it can move its adopters beyond what traditional services were able to offer, and drive new competition at an industry level. In a digital age, where customer convenience and value expectations have rapidly increased, RPA is helping businesses to win in the race to digitisation.

Notes

1 'Get ready for robots', EY (15 May 2016), www.ey.com/en_cz/financial-services-emeia/get-ready-for-robots.
2 'Robots aren't as smart as you think', Angelica Lim, *MIT Technology Review* (2 November 2017), www.technologyreview.com/s/609223/robots-arent-as-smart-as-you-think/.
3 IEEE: Institute of Electrical and Electronics Engineers.
4 *IEEE Guide for Terms and Concepts in Intelligent Process Automation,* 2755-2017, available at https://ieeexplore.ieee.org/document/8070671, italics added.
5 'AGI has been delayed', Rodney Brooks, blog post, *Robots, AI, and Other Stuff* (17 May 2019), https://rodneybrooks.com/agi-has-been-delayed/.
6 Martin Ford, *Rise of the Robots: Technology and the Threat of a Jobless Future*, Basic Books, 2015.
7 'The opportunities in enterprise RPA', Howard Chau, *VentureBeat* (28 July 2019), https://venturebeat.com/2019/07/28/the-opportunities-in-enterprise-rpa/.
8 API: Application Programmatic Interface – an interface that defines interactions between software applications.
9 Eberhard Wolff and Sven Johann, 'Technical debt', *IEEE Software*, Vol. 32, Issue 4 (July 2015), https://ieeexplore.ieee.org/document/7140698.
10 Rosina Moreno and Jordi Suriñach, 'Innovation adoption and productivity growth: Evidence for Europe', working paper, 2014, www.ub.edu/irea/working_papers/2014/201413.pdf.
11 Jacques Bughin and Nicolas van Zeebroeck, 'The right response to digital disruption', *MIT Sloan Management Review* (April 2017), https://sloanreview.mit.edu/article/the-right-response-to-digital-disruption/.
12 'International trends in RPA adoption', Tejus Venkatesh, *UiPath* (6 March 2022), https://rpaconferences.com/assets/pdfs/Tejus-Venkatesh.pdf.
13 'AA Ireland reduce missed webchats by 81% via new chatbots', Clare Ruel, *Insurance Times* (22 October 2019), www.insurancetimes.co.uk/news/aa-ireland-reduce-missed-webchats-by-81-via-new-chatbots/1431669.article.
14 'AXA's admin bots save $250k in six months', Scott Carey, *Computerworld UK* (6 February 2019), www2.computerworld.co.nz/article/657141/axa-admin-bots-save-250k-six-months/.
15 'John Lewis Partnership sells its automation to business', Karl Flinders, *Computer Weekly* (2 October 2019), www.computerweekly.com/news/252471719/John-Lewis-Partnership-sells-its-automation-to-business.
16 'How to tame "Automation Sprawl"', Thomas H. Davenport, *Harvard Business Review* (19 July 2019), https://hbr.org/2019/07/how-to-tame-automation-sprawl.
17 AXELOS, '3. The four dimensions of service management', in *ITIL Foundation, ITIL 4 edition*, 2020. TSO (The Stationery Office).
18 'Get ready for robots', EY (15 May 2016), www.ey.com/en_cz/financial-services-emeia/get-ready-for-robots.
19 Leslie P. Willcocks, John Hindle and Mary C. Lacity, *Becoming Strategic with Robotic Process Automation*, SB Publishing, 2019.
20 'RPA deployment will fail without a strong operating model', Capgemini (6 November 2018), www.capgemini.com/dk-en/2018/11/rpa-deployment-will-fail-without-a-strong-operating-model/.

21 'What process mining is, and why companies should do it', Thomas H. Davenport and Andrew Spanyi, *Harvard Business Review* (23 April 2019), https://hbr.org/2019/04/what-process-mining-is-and-why-companies-should-do-it.

22 'What is OKR? A goal-setting framework for thinking big', Sarah K. White, CIO (4 September 2018), www.cio.com/article/3302036/okr-objectives-and-key-results-defined.html.

23 Sorin Anagnoste, 'Setting up a robotic process automation center of excellence', *Business Management Dynamics in the Knowledge Economy Journal*, Vol. 6, Issue 2 (January 2013), 307–322, www.researchgate.net/publication/326141509_Setting_Up_a_Robotic_Process_Automation_Center_of_Excellence.

24 Gedas Baranauskas, 'Changing patterns in process management and improvement: Using RPA and RDA in non-manufacturing organizations', *Business European Scientific Journal* (30 September 2018), https://pdfs.semanticscholar.org/b2bd/ca48c6c00aa925a499c156a04a1ed45ab5c8.pdf.

25 Leslie P. Willcocks, John Hindle and Mary C. Lacity, *Becoming Strategic with Robotic Process Automation*, SB Publishing, 2019.

26 'How to tame "Automation Sprawl"', Thomas H. Davenport, *Harvard Business Review* (19 July 2019), https://hbr.org/2019/07/how-to-tame-automation-sprawl.

27 Venkat Venkatraman, *The Digital Matrix: New Rules for Business Transformation through Technology*, LifeTree Media, 2017.

28 Phil Fersht, Elena Christopher, Miriam Deasy and Saurabh Gupta, *The New RPA Manifesto*, HFS Research Ltd. (November 2019), www.horsesforsources.com/storage/app/media/2021/RS_1911_HFS-RPA-Manifesto.pdf, p. 2, para. 2; italics added.

29 'Beyond automation: Strategies for remaining gainfully employed in an era of very smart machines', Thomas H. Davenport and Julia Kirby, *Harvard Business Review* (June 2015), https://hbr.org/2015/06/beyond-automation.

30 'Fourth Industrial Revolution', World Economic Forum (6 March 2022), www.weforum.org/focus/fourth-industrial-revolution.

6 Climate Change, Pandemics and Artificial Intelligence

Ruth Taplin and Alojzy Z. Nowak

Artificial Intelligence (AI) can be used to assist in climate change mitigation which is one of the most pressing problems globally. In fact, the Intergovernmental Panel on Climate Change (IPCC) of the United Nations (UN) in February 2022 produced a report stating that more than 40% of the world's population are vulnerable to the effects of climate change and both humans and nature are reaching the point of inability to adapt to such dramatic change.[1] As climate change is negatively affecting countries' economies and societies, AI can be used to good effect to lower carbon emissions and provide data to assist in ending severe air pollution, flooding and land degradation and assist with the needs of rural communities – to warn of plagues of locusts, for example, before they arrive to devour and destroy crops. AI is not only a rapidly developing technology; it has the flexibility to move much more quickly to produce datasets than humans who can then devise rapid solutions.

A recent example is using satellite tracking to understand, through monitoring carbon emissions, where severe air pollution is emanating from. This is particularly important for countries such as India which are trying to move away from highly polluting coal-fired power plants. Such monitoring can be used to understand other sources of air pollution in industrial and electricity plants, and the resulting data can then be used to levy taxes on the heaviest polluters or convince financial backers and the government to finance lesser polluting companies and plants.

Human activity, mainly through intensive use of fossil fuels to run industries, the use of petroleum-derived products such as plastics, industrial-scale livestock production for excessive amounts of meat, estimated at 80 billion sentient creatures being killed for meat a year globally, coupled with reliance on gas and oil for energy requirements have led to unprecedented carbon emissions. These heat-trapping gases, termed greenhouse gases, are at a higher level than in the last 800,000 years of planet earth's existence.

Carbon dioxide emissions have increased by a third since the Industrial Revolution, warming the planet so rapidly that nature cannot change in parallel time. Previously, the environment had thousands of years to evolve to deal with, for example, volcanic eruptions which affected the climate.[2] Therefore,

DOI: 10.4324/9780367857561-6

solutions to these climate problems must be accomplished quickly and that is why AI can offer the most effective and efficient solutions rapidly at pace.

Electricity Systems

Because it is data-driven, AI is and will play a key role in climate change mitigation through a number of key energy systems. As the chapter in my (Taplin's) last book dealing with Cyber Risk and smart grids described,[3] AI has been radically changing energy systems to create electricity-based smart grids that enable low carbon electricity. There are two types of electricity systems that can produce low carbon emissions with the help of machine learning (ML). (It is probably more accurate to use ML here because many experts on AI agree that it is ML which can be used more readily to change systems such as smart grids through reinforced learning AI methods). One is a variable system such as energy derived from the sun or wind and the other is a controllable system such as nuclear or geothermal energy which can be turned on or off.

The bulk of electricity is delivered to consumers through an electric grid where every second of energy consumed is met by power generated. The problem with this system is that power generation is assured largely through carbon-emitting buffer systems known as spinning reserves provided by coal and natural gas plants. To reduce carbon emissions to a level that will mitigate climate change, there is a movement towards new energy storage technologies including batteries, and pumped hydro water systems. Power to gas systems (using hydrogen and ammonia) will also be used, but if operated incorrectly, will cause the same problems with carbon emissions as fossil fuels.

AI-based technologies will be used to lower carbon emissions through shaping forecasting, scheduling and control to create variable electric systems that are flexible in relation to demand.

To assist with fluctuations in electricity generation and demand, scheduling must be made in real time to reduce reliance on polluting standby plants. The basic forecasting techniques which most electric grids rely on at present are not accurate enough and do not operate in real time which AI techniques use to allow efficient forecasting with certainty. AI techniques with their advanced algorithms will be able to improve the accuracy of climate modelling and weather forecasting for solar, wind and water derived from rivers which are driven sources of energy to power electric grids.

At present electric grid system operators use slow and complex scheduling systems called scheduling and dispatch. As scheduling moves to even more complex systems based on increased storage, variable generators and flexible demand, AI can progress a more rapid use of power optimisation systems, with improved quality and speed.

For example, dynamic scheduling and safe reinforcement systems can be utilised to balance the electric grid in real time. Another way in which AI can be used is to decentralise scheduling and dispatch by creating local algorithms for storage and price. For example, AI using real-time measures provides an accurate picture of cost which allows the user to adjust the system to also lower carbon emissions.

AI can be used to create storage facilities for solar energy using new materials; improvements in lithium battery storage can also be achieved using AI-based modelling.[4]

AI and Transportation

Freight and passenger transportation accounts for more than half of carbon emissions related to air travel, vehicles, rail, ships and boats. This sector accounts for more than a third of global carbon emissions. This sector is more difficult to decarbonise because of the density of the fuels used for vehicles and airplanes. Transport activity needs to be reduced to mitigate climate change as well as shifting to lower-carbon emitting types of transportation such as rail, and using alternatives to fuels and electrification.

AI can make a great difference in reducing carbon emissions by enabling road/motorway infrastructure to optimise routing through sensors. Sensors can provide patterns of travel based on accurate and large amounts of data which can be fed into policies for efficient travel infrastructure building; sensors can also be used in vehicle engineering to reduce carbon output. AI can fill the gaps in precise data on vehicle, pedestrian and cyclist counts that are needed to understand traffic flow and road use.

Electric vehicle technologies are viewed as the most important methods to reduce polluting emissions. AI is used in the research and development of batteries by, for example, predicting the state of degradation and length of use of the battery. As electric vehicles are used for short periods during the day and night, they can be used as energy storage units that can feed energy to electric smart grids through machine learning reinforcement techniques. AI can also be used for the development of alternative fuels, as in the storage of electrofuels and hydrocarbons which can be stored at a lower cost than electricity.

Using AI to make accurate predictions of arrival times for passenger buses (displayed on screens or mobiles) or freight so that rail services can be waiting when deliveries of goods are made from the docks can all lower carbon emissions through making transportation systems efficient. AI can also, through sensors, predict rail track degradation to keep the low-carbon rail networks running smoothly.[5]

Industrial Organisation

The global industrial sector, including manufacturing, building materials and logistics, contributes heavily to the production of greenhouse gases. However, it is easier for ML to deal with climate mitigation in this sector, because the sector is data driven. ML has the potential to reduce carbon emissions by improving the quality of production, streamlining supply chains, optimising heating and cooling systems, and predicting machine breakdowns. Yet the use of ML will not have the desired effect unless industry has the incentives, for example, to choose clean electricity over fossil fuels.

Another area in which the use of ML can be increased is predicting with accuracy the supply and demand for goods. Many industries inaccurately overproduce goods which are then wasted.

This includes the fashion industry which manufactures high amounts of waste that produce greenhouse gases through the disposal of unwanted garments because fashions are not predicted correctly.

Food waste is also astronomical with 1.3 billion metric tons of food wasted globally through inadequate storage, estimating supply and demand for certain food products incorrectly, not taking into account time spans to prevent spoiling of certain foods and inadequate refrigeration. ML can be used through sensors to detect when food is about to spoil or when items of food stored with fresher produce will ruin the entire shipment, so it can be quickly removed.

Building materials are another source of high carbon emissions. ML can mitigate such emissions through its use in developing structural products, such as cement, that use less raw materials. 3D printing can be used to provide models for unusual shapes; this would lessen the amounts of materials needed compared to using a traditional cast or mould.

ML can also be used to predict chemical reactions, so in the case of ammonia which is used in fertilisers, ML could be used to develop cleaner ammonia production through the electrochemical analysis of lower temperatures for its production.

ML techniques such as image recognition and time delay neural networks can be used to adapt high energy-use systems to lower-use ones. For example, DeepMind was used through reinforcement learning to optimise cooling centres for Google's internal servers by predicting and optimising the power usage effectiveness, thus lowering both carbon emissions and cooling costs.

ML can also be used as a tool for predictive maintenance, allowing, for example, a factory to test out a new piece of code before it is uploaded by the factory. Through modelling ML can predict wear and tear on factory equipment and through sensors detect leaks, for example, before they become an expensive and dangerous problem that increases emissions.[6]

Agriculture and Forestry

Both industrial-scale agriculture and livestock are contributing at least a third of carbon emissions and the latter even more so with the production of methane gases. While destructive agricultural practices such as draining peat bogs and wetlands, which hold vast amounts of carbon, mean that excessive carbon is released into the atmosphere, methane is an even greater emitter of greenhouse gases. Reversal of these tendencies can mitigate the release of carbon. This means restoring wetlands, peat bogs and stopping deforestation for agriculture and livestock production. Industrial-scale livestock production of 60 billion sentient living creatures causes extraordinary amounts of methane gas and the solution is to stop livestock 'farming', by vastly reducing this industry and turning to alternatives to meat eating; this is happening through education and the increasingly high cost of producing meat.

ML can be used effectively to produce a form of precision agriculture which could stop carbon release from the soil and assure a greater crop yield, thereby mitigating the need for deforestation. ML can predict the risk of fire, monitor emissions and the health of forests. There are robotic tools such as RIPPA – under development at the University of Sydney, Australia – that is equipped with a hyperspectral camera and can carry out mechanical weeding, vacuuming of pests and targeted pesticide application. It can cover five acres per day and bring in large datasets while running on solar energy. Other smart robotic devices are intelligent irrigation systems that decrease water usage and prevent water-borne pests. ML can also assist in disease and pest detection and soil sensing. Through prediction, ML can be used to decide what crops to plant and likely crop yield; there are also models that can predict crop demand to reduce wastage and target consumer choice more effectively. Unmanned Aerial Vehicles (UAVs) with hyperspectral cameras can be used for all these tasks.

ML has been applied through remote sensing data to give estimations of the thickness of peat and assess the carbon stock of tropical peatlands. Peatland mapping can be furthered through satellite applications and using low-cost precise monitoring tools which could also predict the risk of fire.

An example from rural Africa demonstrates how effective AI apps can be for small-scale farmers. In Zambia, there are more than 22,000 small landowning farmers who use an AI-backed mobile application (app) called AgriPredict to obtain instant information concerning weather patterns and plant diseases as well as selling their produce on the platform. With these platforms and hitherto unobtainable expert information from agriculturists and meteorologists, these farmers can now, using their affordable mobile phones, run their farms more efficiently and at a reduced cost.

Mobile phones are available cheaply and are the main source of telecommunications in rural areas in Africa and India – this has improved the chances of rural smallholders to succeed.

Prior to this, agricultural experts had to visit remote rural sites to advise, for example, on the problems with crop production. Such advice was costly, rare and hard to access so AI technology has helped small-scale rural farmers to feed themselves and local communities with nutritious plant-based food without migrating to overcrowded urban areas where they often do not possess the skills to succeed.[7]

Forests

Trees store carbon in above-ground biomass so heights and types of trees provide a good indication of the amount of carbon stock. As many areas of forest are closed to UAVs, ML through satellite imagery can be used to predict or give an informed estimate of carbon stock in forests globally. AI can assist afforestation which is the planting of trees through automated planning which decides the best places to have mass tree planting take place. This process is known as Long-term Uncertain Impact Planting which is the opposite of destructive deforestation and could be used as a means for sequestering CO_2 for the long term.

There is scope for up to 0.9 billion hectares of extra forest cover on the planet. However, mass tree planting must be done with care so as not to obliterate peat bogs and wetlands which also act as carbon sinks. Tree planting must also be carefully monitored, as many such excellent projects are undermined by lack of follow-up care such as tending to the newly planted trees and assessing their growth.

ML can be useful in this process by locating appropriate sites in which to plant trees, assessing weed density and monitoring the health of trees. There are many start-up companies in the United States such as DroneSeed27 and Engineering26 that are assisting with these processes, such as planting seeds more quickly and at a lower cost than previously and managing forest fires. Massive forest fires like those occurring in California release great amounts of carbon into the atmosphere, unlike small natural fires that have occurred for millennia in cycles to assist in clearing low-lying growth for sustained growth. ML can assist through predicting the amount of water in tree canopies and identifying which regions are more prone to fire. Reinforced learning can provide spatial data, assisting firefighters with decisions on when to allow fires to burn and when to stop them and when to carry out controlled burning.

The logging industry is particularly destructive, especially when it uses clearcutting which refers to indiscriminate felling of trees rather than doing so carefully and selectively. Through sensory imaging, ML can differentiate between clearcutting and selective tree felling.

Old smartphones powered by solar panels can be installed in forests to detect, through reinforced ML, the sound of chainsaws – to alert law enforcers that illegal logging is taking place. ML tools are also becoming increasingly used by foresters to determine where to harvest, fertilise, build roads and fire breaks.[8]

Tracking Carbon: A UN Goal

Carbon Tracker, a non-profit think tank, based in the City of London, that makes finance recommendations for companies to transition from fossil fuels to sustainable energy is using ML through satellite imagery to find the sources of carbon-emitting power plants that are contributing to air pollution. With a grant from Google, Carbon Tracker, the first of its kind, will use AI technologies to detect carbon-emitting power plants globally and provide the results to the public, so that power plants can be held accountable for their emissions. In collaboration with WattTime and the World Resource Institute

(WRI), organisations which have experience in using AI techniques to transition towards clean energy, Carbon Tracker will use the most advanced image-processing algorithms to detect emissions from space, enabling them to track power plant emissions globally. Sensors that operate from a variety of wavelengths using algorithms produce data which can accurately check power plant emissions by reading thermal infrared signals from heat generated both by smoke stacks and the intake of water for cooling purposes. Visual spectrum recognition will also be used to confirm that a power plant is emitting smoke. Carbon Tracker can be used ultimately to make those companies which are emitting the most carbon to take responsibility for it through, for example, extra fines and taxes.[9]

Locust Tracking Using AI Apps

AI apps can not only assist with carbon tracking, but in tracking devastating pest invasions that ruin smallholding farmers' livelihoods and can cause famine. In rural regions of India, it is difficult enough sometimes surviving with little, if any, outside support and being isolated in remote areas. This is becoming increasingly the case, with the effects of climate change meaning that rainy seasons, such as monsoons in India, which are vital for crop success do not arrive in time, or conditions of drought mean that unexpected pests increase their numbers with devastating consequences.

A recent example that swept East Africa involved huge numbers of locusts. The apps developed to deal with this crisis can be transferred to other regions in the world that are suffering from local agricultural disasters induced by climate change.

In 2020, billions of locusts descended on a number of East African nations. A locust attack of this intensity had not been seen since the 1950s and this time the size of the swarms was being blamed on unusually warm weather brought about by climate change. Nineteen million farmers had their livelihoods affected in Kenya, Ethiopia and Somalia, countries which had not experienced such swarms in 70 and 30 years respectively.

The impact of the locust attack could have been much greater if a groundbreaking technology-driven operation had not been employed. It was a collective effort that created a locust database as a simple smartphone app; this replaced an earlier tablet-based locust tracker app that was on a tablet which is no longer manufactured. The other problem was that even if the original tracker programme – which operated through transmission of satellite imagery to agricultural experts who recommend strategies to combat such locust swarms – had still worked, Kenya and the other affected countries did not have enough agricultural experts.

PlantVillage and Climate Change Technology

PlantVillage is a Kenyan non-profit organisation that assists farmers in dealing technologically with agricultural-based disasters induced by climate change.

It was in the forefront of lessening the impact of the 2020 locust swarms. The damage potential of these swarms was so great as to interfere with food security in the region; the crisis therefore drew in an interdisciplinary group of experts from around the world. Dr David Hughes, a US entomologist from Pennsylvania State University, thought it would be a waste of experience not to find solutions that would make it possible to deal with future locust swarms and other pest outbreaks around the world. He noted that the magnitude of locusts could be directly linked to climate change with the increase in cyclonic activity.

The 2020 plague of locusts began in 2018 in a remote part of Saudi Arabia which had experienced two major cyclones; these produced so much rain that it caused an 8,000-fold multiplication of locusts in the desert region. Winds then pushed the locusts into the Horn of Africa by the middle of 2019, and a wet Autumn further increased their numbers. This was followed by an unusually wet December in Somalia that turned the swarm of locusts into a state of emergency in the region.

Working with Keith Cressman, a senior locust forecasting officer at the Food and Agriculture Organization of the United Nations (UNFAO), it was thought it would be more effective to create a mobile smartphone app for all involved to collect locust data than to return to creating software for new tablets. He asked David Hughes who had already created a smartphone app with UNFAO to combat an influx of the destructive armyworm crop pests and implemented it through PlantVillage, the organisation he created.

Given the fact that such a large swarm of locusts can consume enough food to feed 13,000 people, producing a tracking app was an urgent task. The original PlantVillage app uses both AI and ML to assist farmers with their crop production in more than 60 countries. Using this as a blueprint, Dr Hughes and his colleagues created a new app named eLocust3m within a month. The app presents photos of locusts at different stages of their lifecycles. This assists users' diagnosis of what they view in the field. GPS coordinates are automatically recorded and AI algorithms are used to double-check photos submitted with each entry. Garmin International, a well-known GPS provider, also supported another programme that worked on satellite-transmitting devices. "The app is really easy to use", said Ms Melodine Jeptoo of PlantVillage. Last year, she recruited and trained locust trackers in four devastated Kenyan regions: "We had scouts who were 40- to 50-year-old elders, and even they were able to use it."

The locust tracker app and other measures made effective by AI methods have reaped important results since February 2020 in East Africa, with an estimated US$1.5 billion of commercial produce and 34 million livelihoods saved. In 2022 drought and adverse conditions due to climate change are affecting nations of East Africa and there is more need than ever for AI solutions.

Dr Hughes has been working with the National Oceanic and Atmospheric Administration (NOAA) and the Massachusetts Institute of Technology (MIT) to use locust reports to build AI big data models that will predict future plagues. Insights provided by these models would allow countries to implement pre-emptive control strategies which are less damaging to the environment than pesticides.

Dr Hughes noted that these approaches could also be used to combat other climate- induced disasters, such as pests, floods and droughts. The fight against locusts demonstrated that crowdsourcing can be used to fund AI projects that can support hundreds of millions of people affected by the potentially deadly effects of climate change.[10]

AI Fighting Methane Emissions and Disease

AI is being increasingly used to produce alternatives to dairy, meat and fish products and to fight disease or find the source of the disease.

Imagindairy, which is an Israeli technology company start-up, is producing milk proteins including whey and casein from plant micro-organisms like fungi. Dr Eyal Afergan who is the company's Chief Executive and Co-Founder, states that AI is used to programme the DNA that is inserted into the fungi which is used to grow the protein. The proteins produced by Imagindairy are identical in texture and taste to those found in cow's milk. The bonuses of Imagindairy milk are that it is without lactose, sugar and fats and reduces methane emissions as no cows are involved in the production cycle.

Imagindairy is planning to supply dairy companies with milk protein powder that can then be used to make milk, cheeses and yogurts. As two-thirds of people in the world are lactose intolerant, a non-dairy milk such as that produced by Imagindairy holds further benefits.

It may also have another indirect benefit by greatly reducing the amount of soya milk, since soya beans are grown in abundance in Brazil, largely for the East Asian market; this results in unnecessary deforestation of the rainforests which act as carbon sinks, absorbing vast amounts of carbon emissions in the forest canopies, as described earlier in this chapter.[11]

AI and COVID-19

One major aspect of the COVID-19 (coronavirus) pandemic that is a harbinger of change concerns the food choices that we make in the future; this will ultimately determine the survival of both humans and wildlife. Cheap food becomes very costly if it destroys people, animals and planet earth.[12] As mentioned above in relation to Imagindairy, AI can be used to produce alternatives to meat and dairy that will make a major contribution to mitigating climate change by removing methane emissions, just as turning away from fossil fuels and embracing renewables will substantially reduce carbon emissions. This may also solve potential deadly pandemics such as COVID-19 because inexpensive alternatives to meat and dairy, which can be imbued with the flavours of culturally accepted food, will mean that the risk of eating wildlife and bush meat will not be necessary. The latter is, and has been, the cause of local and historically recent global pandemics that derive from so-called wet markets and meat found in the bush which have given rise to Ebola, HIV, SARS and COVID-19. The vital and socially important role that AI can play in pinpointing the likely origin of such pandemics is discussed later in the chapter.

Antibiotic-Resistant Strains

In reality, we eat food such as 'burgers' which cost far in excess of the cheap food offered at a fast-food outlet. To produce the meat content is very costly in terms of water, fertiliser, antibiotics and the lives of those who died at meatpacking plants in the United States, for example, because of the coronavirus. The cheap burger can only be profitable for the retailer if it sells it in large quantities which destroys the planet's resources and human health in terms of ever-increasing obesity, diabetes, heart disease and the erosion of the effectiveness of antibiotics.

According to Dr Tedros Adhanom Ghebreyesus, Director of WHO (World Health Organization), COVID-19 has contributed to the overuse of antibiotics which is leading to severe antimicrobial resistance. He noted that increased use of antibiotics and antimicrobial medicines in both humans and animals is rendering their medicinal use ineffective in disease and death prevention globally. This is because microbes are becoming increasingly drug- resistant.

Industrial-scale farming, as a matter of course, uses antibiotics in feed to prevent outbreaks such as salmonella. It is the unhealthy cramming together of poultry, for example, in both exceptionally cruel and unhygienic conditions that causes disease in these short-lived living "commodities" which requires overuse of antibiotics which humans then ingest, weakening their immune systems. COVID-19 (and previous pandemics emanating from China) and Ebola in Africa have direct connections to humans eating wildlife prepared in horrifically cruel and unhygienic places called wet markets.

Real Price of 'Cheap' Food

Fast 'cheap' food based on meat is extremely costly to the environment. The hidden costs of a McDonald's Big Mac, for example, are in reality £150.00 per burger. This is according to the Indian Centre for Science and the Environment. The price is so high because of all the hidden costs that we and fast-food companies do not pay, such as deforestation in the Amazon for cattle farming which kills the lungs of the planet, methane from cows that heats the planet, overuse of antibiotics that affects adversely our immune systems, production of toxic slurry that poisons the soil and water resources on which we depend for nutritious food production.

True Value Food

Today, British families spend on average 8.2% of their income on food while in the United States, it is 6.4%. In 1950s Britain, by comparison, families spent

30% to 50% of their income on food. Cheaper now, but low in value, the food produced by intensive farming today leaves us malnourished, relying on added vitamin supplements, obese as well as diabetic, relying on unhealthy diet fads in addition to more unhealthy drugs to combat these preventable conditions. Carrots, for example, lost 75% of their copper and magnesium between 1940 and 1991. Crops are dosed with chemicals in the form of fertilisers and pesticides. Crop rotation is not practised so soil becomes infertile, providing even less nutrients to our food. Deforestation, including pulling out hedgerows or other sanctuaries for wildlife, has resulted in an alarming decline in wildlife globally. Birds in Britain which used to eat aphids and other insects that kill plants have declined in number by 55% since 1970. Plant-based foods are not only affordable but nutritious, as beans, vegetables and fruit use relatively little water or artificial fertilisers, no antibiotics and can be grown on the edge of cities, on allotments and can avoid pesticides through tried-and-tested mixed planting methods as well as using crop rotation to stop soil exhaustion. Value is an essential concept, but perhaps one of the most difficult to define as it can be both subjective and subject to many different variables. In relation to food, it is not good value to pay less for a food that is produced at a high cost to every resource we depend on, such as water, soil, health, civilised behaviour, biodiversity and social need. Therefore, good value should be food that is produced at the highest standard, is nutritious, does not kill animals (sentient beings), is produced without cruelty, gives back to the soil nutrients and does not pollute the air we breathe and the water we drink. Organic farming can be inexpensive in that what is consumed is nutritious food without waste while replenishing the soil. Wildlife will flourish and re-assert their role as pollinators (bees are in decline) and eaters of insects (many bird species are in decline), rather than using chemical-based pesticides which are poisonous to all. AI robots can be used to assess the quality of the soil, weed efficiently without pesticides being used and predict the best crops to plant for maximum yield.

Pandemics and Eating of Wildlife as 'Cheap' Food

What is definitely not good value is for governments around the world to turn a blind eye to the killing and eating of wildlife and destruction of their habitats. COVID-19 and the long list of pandemics before it emanating from China are the result of indiscriminate, brutal killing of wildlife in the most unhygienic circumstances to allow large populations to obtain free or cheap food. The price in real terms of these dubious practices is very high as has been witnessed in worsening pandemics, increasing in deadliness and in mutations that are difficult to combat. Wild animals harbour viruses that can be transmitted to humans, mutating and becoming even more deadly. This is especially true, for example, when civet cats that have been eating bats are transported to wet markets. The civet cats are ill from maltreatment, stacked in cages one upon another so bodily fluids drip onto others, and kept in highly unhygienic conditions until sold to be brutally killed for consumption.

The dangers of eating wildlife from wet markets in China have long been known. Dr Shi Zhengli, is well known throughout the world and in China for her research on virus transmission from bats. She was part of the team that traced the previous SARS (severe acute respiratory syndrome) pandemic to horseshoe bats eaten by civet cats that had been slaughtered cruelly and unhygienically in Chinese wet markets.

Dr Shi, the Deputy Director of the Wuhan Institute of Virology, had warned about the dangers of future epidemics from bat-borne viruses. Her Institute knew through gene-sequencing and ML tests as early as 2 January 2020 about the new coronavirus and its origins. However, the Director of the Institute sent emails warning staff not to disclose any information about the discovery. Eight doctors in China tried to bring to the attention of the populace the dangers of what was to become known as COVID-19, but were arrested for spreading 'rumours' which cost some of them their lives from the novel coronavirus.

As the Chinese government could no longer suppress the explosion of the deadliest pandemic to originate from China through the eating of wildlife from unhygienic wet markets, conspiracy theories began to explode as well, with one in particular the most misleading. It suggested that the novel coronavirus emanated from Dr Shi's laboratory which was an unlikely scenario as the Wuhan Institute, built with the assistance of the French, has the highest bio-security measures in place.

As the disease spread and deaths mounted in China, one report appeared in *The Beijing News* identifying a researcher at the Institute as 'patient zero' – the first person to be infected; while it was those who consumed wildlife meat at the Wuhan wet market who were the first to contract the virus.

Shi was subjected to severe attacks on social media as the 'mother of the devil'. Yet she issued a strident denial on her WeChat social media account, stating that the new virus was "nature punishing the human race for keeping uncivilised living habits".

"I swear with my life – [the virus] has nothing to do with the lab", she declared, telling those spreading false rumours to "shut their stinking mouths".

Dr Shi has worked alongside many of the world's top experts on infectious diseases. "She is a superb scientist and [a] very nice person", said James LeDuc, Director of the Galveston National Laboratory, a high-security biocontainment centre in Texas. "She has been very open and collaborative for the decade I've worked with her."

AI Can Assist in Stopping Illegal Wildlife Consumption

"We can't be indifferent anymore!" President Xi Jinping of China fumed at senior officials referring to the public health risks of eating wildlife. On February 24 2020, the 13th National People's Congress issued a declaration, "Comprehensively Prohibiting the Illegal Trade of Wild Animals, Eliminating the Bad Habits of Wild Animal Consumption and Protecting the Health and Safety of the People". This and an earlier ban on wildlife markets were direct responses to concerns that the emerging novel coronavirus, which is thought to have originated in bats, had been transmitted to humans through an unhygienic, so-called wet market which cruelly sells live wildlife, in Wuhan, a city in Central China, for slaughter and consumption. The purported closing of wet markets to stop new transmissions of deadly pandemic viruses from wild animals to humans and the banning of the wildlife trade has been in the Chinese law books for 30 years. The problem, using the Pangolin, a gentle anteater with scales as an example, is that actual enforcement is virtually non-existent and not deemed a priority. In fact, the wildlife trading bans in China have not stopped the Pangolin from becoming the most trafficked mammal in the world. Pangolins are almost extinct now in China according to a Chinese organisation, the China Biodiversity Conservation and Green Development Group. They are also declining in Southeast Asia, especially in Malaysia and Indonesia, to such an extent that poachers are now turning to a supply from West Africa via Nigeria. According to the conservation group TRAFFIC, a global wild-life trade monitoring network, 90,000 Pangolins were smuggled into China between 2016 and 2017 despite a commercial trading ban on all eight species of Pangolin under the 1973 Convention on International Trade in Endangered Species [CITES] of Wild Fauna and Flora to which China was a signatory.

Over the past 20 years, Malaysia has had an 80% decline in Pangolin numbers while the Philippines and India have had a 50% decline.

AI machines are now being used to monitor the illegal trade in wildlife, much of which is headed for wet markets which fuel epidemics like COVID-19. AI is being used to find patterns in illegal trafficking practices – for example, from Nigeria to China or Vietnam – allowing law enforcers to narrow the search for the likely suspects engaged in this illegal activity who are often criminal gangs.

Dangers of Traditional Medicine

The demand for Pangolin flesh and scales is driven by Chinese demand, ranging from senior Chinese officials who gain a high status from serving Pangolin hotpot to guests to Chinese 'medicine' companies that use Pangolin scales for dubious medical treatments.

Chinese bans on the consumption or trade of wildlife are undermined by legal loopholes that allow for such trade online and black markets for 'medicinal' or 'research' uses – thus, the wildlife trade continues to flourish. Yet this is despite warnings in historical Chinese medical literature warning of the poisonous dangers of eating wildlife including Pangolins. The scales of the beleaguered Pangolin are believed to have health-giving properties. In a survey from 2015, it emerged that 70% of Chinese respondents believed that consuming Pangolin could cure rheumatism and skin diseases and heal wounds. People hold some of these beliefs thinking they are rooted in traditional Chinese cuisine and medicine, but they are located in superstition and misinformation.

Quite to the contrary, the meat of Pangolins was believed to cause ailments, rather than curing them. It tastes bitter and was thought to be poisonous. The *Beiji Qianjin Yaofang* (备急千金要方, a collection of prescriptions compiled by Sun Simiao, an alchemist of the Tang dynasty), advised in the year 652: "There are lurking ailments in our stomachs. Don't eat the meat of pangolins, because it may trigger them and harm us". The *Bencao Gangmu* (本草纲目, *Compendium of Materia Medica*), the Chinese medicine and cuisine capstone by herbalist, naturalist and physician Li Shizhen (1518–1593) warned that people who eat pangolin "may contract chronic diarrhoea, and then go into convulsion and get a fever".

Ancient literature also warned against eating any number of other wild animals, including snakes, badgers, boars and the bats that have been linked scientifically in modern times to the transmission of deadly virus-based diseases to humans.

One author of this chapter (Taplin) remembers seeing ancient Han period glazed turquoise pottery which showed how in ancient times, Chinese people shared their homes with ducks and pigs (the worst combination for transmission of deadly viruses).[13] It suggests that living in close proximity with these animals which can harbour viruses that can be transmitted to humans provided a foundation for pandemics to occur to this day. Therefore, eating unhygienically prepared wildlife on a mass scale has increased the risk massively.

The knowledge of the ancients is ignored in China in favour of status, money, superstition and cheap food in the case of the consumption of dogs, cats and commonplace wildlife.

Yet after testing more than 1,000 samples from wild animals, scientists at the South China Agricultural University found the genome sequences of viruses found on Pangolins to be 99% identical to those on coronavirus patients. Luckily, because of COVID-19, there has been a drop in the demand for Pangolin bushmeat from Central African forests which is then sold to China via Nigeria. Conservationists are trying to assist in the situation continuing after the pandemic subsides.

A Question of Enforcement, Determination and AI

Despite the fact that all eight species of Pangolin are protected under international law and three of the four that are native to Asia are included on the Red List of the International Union for Conservation of Nature (IUCN) as critically endangered species, the legal Chinese loophole of exempting protection for species used in traditional Chinese medicine has been highly undermining.

Therefore, news of their delisting from the traditional Chinese medicine (TCM) pharmacopoeia, reported by *China's Health Times* newspaper, is very welcome. This comes after the country's State Forestry and Grassland Administration (SFGA) raised the protected status of Pangolins to the highest level, with immediate effect, on a par with pandas, tigers and now jaguars from Latin America. Zhou Jinfeng, Secretary General of the China Biodiversity Conservation and Green Development Foundation (CBCGDF), who has long pushed for better protection of Pangolins and for stopping the use of their scales, stated "Our continuous efforts for several years have not been in vain".

Oddly enough, it has been the coronavirus that has drawn attention to the Chinese flouting of international law concerning the illicit trade in wildlife for food consumption and TCM. Prior to COVID-19, the scale of the abuse of wildlife in secret wildlife 'farms' in China and fuelling of the demand in Nigeria, Malaysia, Indonesia and Vietnam was not understood. In February 2020, China's National People's Congress instituted a ban on the consumption of meat from wild animals, although uncertainty remains as to what wildlife will still be allowed for use in TCM and the fur and leather industries.

Although the moves by China's government to ban wildlife wet markets, wildlife farms and trade in wild animals such as Pangolins for consumption and medicine is a step in the right direction, details of which wild animals (or preferably all) will be subject to the ban and how it will be enforced on the black market needs to be addressed. A spokesperson for animal welfare campaign group, World Animal Protection, told the news service that the protection should be extended to all wild animals, "who, like pangolins, are poached from the wild and often placed in squalid, cramped cages, creating a lethal hotbed of disease".

The largest wholesale vegetable and meat market, Xinfadi in Beijing, has been a new source of COVID-19 in China. Virus contamination was found on an unhygienic chopping board used for salmon which caused a panic and cancellation of all salmon imports. It emerged, however, that the salmon was not the source of infection, but 40 samples of virus contamination was found. It is claimed that no wildlife was sold there as in the Wuhan wet market, but if meat was sold, its derivation could be confused especially as the virus is spread through animal-to-human or human-to-human contact. Such viral receptors for spreading coronaviruses come from animal mammals or humans, not fish. Therefore, it is likely that the source of the contamination was derived from meat sold within the market.[14]

COVID-19 has re-emerged with a vengeance in China and the solution has been radical and repressive lockdowns which have led to civil unrest as the populace in hard-hit Shanghai are running out of basic supplies and being sent to poorly equipped makeshift quarantine hospitals, being compared to death camps by some, if testing positive for COVID-19.[15]

The irony is that China is investing huge amounts of resources into developing AI solutions for military and repressive purposes, as outlined in Chapter One, but does not appear to be investing in AI solutions that can cope with COVID-19 and treating those affected. The source of the corona-virus pandemic was the wet market in Wuhan, and wet markets continue to exist. Officials from the World Health Organization (WHO) who were not from China used AI mapping tools to demonstrate that the COVID pandemic did emanate from the Wuhan wet market.

The WHO researchers, using AI techniques, mapped cases from between January and February 2020. They extrapolated from data collected by Chinese researchers from Weibo, a social media app that had created a channel for people with COVID-19 to seek medical help.

The WHO experts led by Dr Michael Worobey had already identified 164 cases of COVID-19 in Wuhan over the course of December 2019, but the cases were marked by fuzzy dots scattered across a featureless map of Wuhan.

Dr Worobey and his colleagues used AI mapping tools to estimate the longitude and latitude locations of 156 of those cases. In a city of 11 million inhabitants, they found that the highest density of the December cases were centred on the Wuhan wet market. In this relatively tiny space, cases included not just people who were initially linked to the market, but others who lived in the vicinity.

Dr Worobey and his colleagues therefore concluded that these AI-produced patterns pointed to the market as the origin of the outbreak. The WHO researchers additionally ran further AI tests which demonstrated that it was extremely unlikely that such a pattern could be produced merely by chance.[16–17]

AI machine learning and techniques are invaluable as shown by the WHO researchers and can be used to prevent future pandemics and mitigate climate change when used wisely by humans.

Notes

1 'Climate Change: IPCC Report Warns of "Irreversible" Impacts of Global Warming', Matt McGrath, BBC News Online (28 February 2022), www.bbc.com/news/science-environment-60525591.
2 'Causes and Effects of Climate Change', Christina Nunez, *National Geographic* (22 January 2019).
3 Ruth Taplin, *Cyber Risk, Intellectual Property Theft and Cyberwarfare: Asia, Europe and the USA* (Abingdon; Routledge, 2021). See Chapter Seven.
4 'Electric Systems', Priya L. Donti, section 2 in 'Tackling Climate Change with Machine Learning' (5 November 2019). This is based on a paper from an AI conference in June 2019 which had a workshop organised by postdoctoral student David Rolnick of the University of Pennsylvania. The paper that resulted was supported by National Science Foundation grant 1803547, then by the Center for Climate and Energy Decision Making through a cooperative agreement between the National Science Foundation and Carnegie Mellon University (SES-00949710), US Department of Energy contract DEFG02-97ER25308, the Natural Sciences and Engineering Research Council of Canada, and the MIT Media Lab Consortium.

5 'Transportation', Lynn H Kaack, section 3 in 'Tackling Climate Change with Machine Learning' (5 November 2019).

6 'Industry', Anna Waldman Brown, section 5 in 'Tackling Climate Change with Machine Learning' (5 November 2019).

7 'How AI Is helping Rural African Farmers', BBC News Online (1 December 2020), www.bbc.com/news/av/business-55138290.

8 'Farms & Forests', Alexandre Lacoste, section 6 in 'Tackling Climate Change with Machine Learning' (5 November 2019).

9 'Carbon Tracker to Measure World's Power Plant Emissions from Space with Support from Google.org', Press Release from Stephano Ambrogi of Carbon Tracker (8 May 2019), https://carbontracker.org/carbon-tracker-to-measure-worlds-power-plant-emissions-from-space-with-support-from-google-org/.

10 'As Locusts Swarmed East Africa, This Tech Helped Squash Them', Rachel Nuwer, *The New York Times* (8 April 2021), www.nytimes.com/2021/04/08/science/locust-swarms-africa.html.

11 'Milk from Israeli Lab Has No Need for Cows', Anshel Pfeffer, *The Times* (18 November 2021), www.thetimes.co.uk/article/milk-from-israeli-lab-has-no-need-for-cows-75q9rq26n.

12 'Cheap Food? Too High a Price!', Ruth Taplin, *Interdisciplinary Journal of Economics and Business Law*, Vol. 9, Issue 3 (2020), pp. 119–128. This article is the reference source also for the next seven sections of the chapter, including quotations.

13 Ruth Taplin, *Decision-Making and Japan: A Study of Corporate Japanese Decision-Making and Its Relevance to Western Companies* (Japan Library, 1995; reprinted by Routledge, 2003 and 2013), pp. 30–36, for discussion of this practice under the Chinese feudal system, by contrast with the practice in England and Europe (in the period after the Black Death) and Japan (which had strict codes of hygiene).

14 'Cheap Food? Too High a Price!', Ruth Taplin, *Interdisciplinary Journal of Economics and Business Law*, Vol. 9, Issue 3 (2020), pp. 119–128.

15 'China Covid: Clashes in Shanghai over Lockdown Evictions', Robin Brant, BBC News Online (15 April 2022), www.bbc.com/news/world-asia-china-61117528.

16 'New Research Points to Wuhan Market as Pandemic Origin', Carl Zimmer and Benjamin Mueller, *The New York Times* (February 27 2022, updated), www.nytimes.com/interactive/2022/02/26/science/covid-virus-wuhan-origins.html.

17 'Shanghai: Censors Try to Block Video about Lockdown Conditions', BBC News Online (23 April 2022), www.bbc.com/news/world-asia-china-61202603.

7 Artificial Intelligence

A Looming Economic and Moral Crisis

Kenneth Friedman

AI → 危機

'AI' is the abbreviation for 'artificial intelligence,' making the left side of the arrow straightforward. The right side may be less so. It is the Chinese character for 'crisis'. That character is composed of two characters, wēi (危) and jī (機/机). The former is readily translated as 'danger,' while the latter has often been translated as 'opportunity' – though it has been argued that this translation is overly optimistic, and that 'inflection point' would better convey the spirit of the term. For the purposes of this discussion, the more common and hopeful translation of jī as 'opportunity' will be adopted.

機/机 – Opportunity

The Industrial Revolution: Great While It Lasted, but Finally Winding Down?

History does not flow smoothly. Progress is typically characterized by fits and starts, often separated by long periods of stagnation or decline.

> The impressive ruins of Roman-era houses and monuments, the number of shipwrecks, the volume of manufactured goods, the level of industrial pollution in ice cores, and the staggering numbers of animal bones from settlements make it clear that Western energy capture was higher in the first century CE than in the eighth or even the thirteenth, but how much higher? Ingenious calculations by economic historians point toward an answer. Robert Allen (2007a) has shown that in 300 CE real wages (which, for the most of the poor in premodern times, closely mirrored energy consumption) in the Western core were comparable to those of southern Europe in the eighteenth-century CE, and Walter Scheidel (2008) has suggested that Roman-era wages were comfortably higher than those in much of medieval Europe. Data gathered by Geof Kron (2005) and Nikola Koepke and Joerg Baten (2005, 2008) indicate that stature changed little between the first and eighteenth centuries, and

DOI: 10.4324/9780367857561-7

Kron (forthcoming[1]) suggests that ancient housing was typically better than that in the richest parts of eighteenth-century Europe.[2]

The late eighteenth and nineteenth centuries made up for nearly two millennia of economic malaise. The key was the development and refinement of the modern steam engine. Even that started slowly. The first modern steam engine (the ancient Egyptians appear to have toyed with the use of steam power), the 'Miner's Friend,' was highly inefficient and had the single limited use of dewatering coal mines.

In the 1770s James Watt separated the condenser from the evaporation cylinder, greatly increasing the efficiency of the machine. Daughter inventions, progenies of the steam engine, spurred British economic growth over a century. Entering the market for cotton cloth, 'spinning jennies' provided far greater efficiency and produced a higher-quality product than the previous hand labor on spinning wheels, increasing cotton production more than one-thousand-fold in 70 years. Related inventions had parallel impacts in other industries, dramatically transforming the British economy, increasing total production while reducing costs.

Over the next century, the fruits of these inventions spread to Europe and North America, where they spawned further inventions and sharply boosted economic growth.

In retrospect, James Watt's improvements to the steam engine and the chain reactions of subsequent inventions that they engendered mark what may be the most important turning point in the roughly 300,000-year history of *homo sapiens*. One might even argue, along the lines of Steven Pinker, that it changed human nature – or at least the way in which that nature was expressed.[3] For it enabled many people to distance themselves from the constant struggle for survival, providing the time and luxury to read novels, stimulating the ability to see the world through the eyes of others, an important source of empathy. A reasonable assurance of survival with a modicum of comfort may provide the room necessary for empathy and morality.

More broadly, the Industrial Revolution changed the economic landscape at many levels. It eliminated broad areas of employment, especially where machine could replace muscle, but created new areas of employment, many of which turned on mental acumen, rather than physical strength and stamina. Socially, increased *per capita* wealth enabled some countries to construct social safety nets to alleviate extreme poverty. (Ironically, some orthodox economists have opposed this on the grounds that redistribution disincentivizes the most productive, arguing that the prospect of poverty can be a potent incentive to work harder and be more productive.[4]) It enabled them to provide public education systems that would eliminate illiteracy. It enabled some to wage war more efficiently and at longer distances, or to create more efficient police states.[5]

Economically, it facilitated the translation of theoretical scientific insights into impressive technological and engineering achievements. This has been of particular importance, as it has been technological breakthroughs that have largely driven economic progress over the past century.[6]

However, the Industrial Revolution is aging, and the growth rate of the industrialized countries has slowed markedly over the past half century, several seeing productivity growth decline by more than half. (A rebound in the 1990s appears to have been short-lived, with U.S. labor productivity growth from 2007 Q4 to 2016 Q3 [most recent] being less than half its average over the previous quarter century.[7]) Moreover, many forecasts extrapolate a continuing, if gradual, decline in productivity growth well into the future. It may be that after two centuries of dramatically increasing productivity and standards of living, the Industrial Revolution has lost its momentum.

This does not augur well. Throughout the past, in both East and West, a common alternative to eras of economic progress has been not stagnation, but economic decline or collapse, often accompanied by increasing violence and failed government as social fabrics have frayed and as relatively legitimate regimes have been displaced by warlords.[8] These have been hellish times for the great majority of people.

In view of not only these considerations, but also the consideration that broad-based economic growth may be a prerequisite, necessary if not sufficient, to our addressing other critical problems that we face, reversing the long-term decline in economic growth is of paramount importance. But how might we achieve that?

'The King is Dead. Long Live the King.' Computers and AI: The Promise of New Growth

The Origin of a Species: The Descent of Artificial Man

Computers do not resemble steam engines in either form or function. But they owe their existence to the steam engine. As just one consideration, electricity is necessary for computers, and despite recent technological advances, solar cells and wind turbines, nearly all electricity is still produced by steam turbines. Early computers, using prodigious amounts of electrical power, required massive amounts of steam. Moreover, computers do not only depend on steam engines, appearing only after the output of steam reached a critical threshold, but like them, they are complex machines designed to increase the productivity of humans.

Similarly, AI, also designed to increase the productivity of humans, is a direct descendant of computers, appearing only when computing power reached a critical threshold.

The Exponential Increase in Computing Power: Moore's Law – A Growth Engine

Archimedes is reputed to have said: 'Give me a place to stand and a lever long enough, and I shall move the world.' Imagine an economy on one side of a fulcrum and exponentially-increasing weight in the form of computer power on even the short side. As the additional weight exceeds the torque of the economy, computer power moves the economy further and faster.

The primary descriptor of this increasing weight of computer power is Moore's Law, an empirical generalization suggested by Gordon Moore in 1965 from just a few years of data. He extrapolated that data to surmise that computing power, the number of transistors in a dense integrated circuit, would continue to double every year for a decade. He later revised that extrapolation to doubling every two years. Remarkably, that growth continued for half a century and is only now beginning to tail off. A cell phone today contains more computational power than the sum of all the world's computers in 1970.

We have witnessed dramatic advances not only in memory, but in speed, size, and expense. The extent of the advances in a relatively short time makes science fiction seem tame.

> The ASCI Red, the first product of the U.S. government's Accelerated Strategic Computing Initiative, was the world's fastest supercomputer when it was introduced in 1996. It cost $55 million to develop and its one hundred cabinets occupied nearly 1,600 square feet of floor space… Designed for calculation-intensive tasks like simulating nuclear tests, ASCI Red was the first computer to score above one teraflop… To reach this speed it used eight hundred kilowatts… By 1997 it had reached 1.8 teraflops.
>
> Nine years later, another computer hit 1.8 teraflops. But instead of simulating nuclear explosions, it was devoted to drawing them and other complex graphics… It did this not for physicists, but for video game players. This computer was the Sony PlayStation 3, which matched the ASCI Red in performance, yet cost about five hundred dollars, took up less than a tenth of a square meter, and drew about two hundred watts.[9]

Talents: An Independent Combinatorial Source of Exponentially Increasing Artificial Intellect

There is an independent driver of AI, one that is combinatorial. AI programs are designed to accomplish particular tasks, t. To achieve this, they often confer some particular Talent, T, on an inanimate object. There are many tasks, t_i, and some of these are complex. And so, there are many talents, T_i. Moreover, it is often possible to combine different Talents, $T_i + T_j$. And these combinations may be worth far more than the individual Talents themselves. The Talent of steering is necessary for an autonomous vehicle. But it would be worth little without the additional Talent of recognizing what is in the road and what the vehicle must avoid.

Suppose there are already n Talents and someone introduces a new Talent, T_{n+1}. This new Talent may be valuable in itself. But it may also be valuable in combination with any other Talent – or group of Talents. There are 2^n-1 possible combinations. With each new Talent, the number of possible combinations doubles. If, on average, each combination has the same economic value – and this is intended as an illustration, a questionable first approximation – then the economic impact of AI doubles with each new Talent, primarily because each new Talent increases the value of the existing Talents.

If the number of Talents is proportional to computing power and *if* that computing power doubles every N years, then the number of Talents doubles every N years. *If* in addition, the average value created by each new Talent (including combinations with previous Talents) does not decline too quickly, then each Talent doubles the economic impact of AI. AI, by itself, would produce not just wealth that is unbounded, but a rate of economic growth that is unbounded.

Remarkable Progress Despite the Winding Down of Moore's Law

The very notion of exponential growth whose rate itself increases exponentially provides grounds for unlimited optimism. Even if Moore's Law 'slows,' the exponential increase in combinations with each new 'Talent' would still be formidable. Respected scholars and astute thinkers have argued intelligently for such optimism,[10] and several Western countries did experience a technology-driven productivity mini-boom for about a decade near the end of the twentieth century.

However, despite the fact that the increase in computer capabilities has far exceeded the expectations of the most optimistic pioneers, there must be limits. Reasoning backwards from the implausibility of the conclusion of an unbounded rate of economic growth, at least one of these *ifs* must be overstated. It may be that all of them are. It is plausible that there is a limit to the number of independent economically significant 'Talents,' and the evidence of the past few years suggests that Moore's Law is slowing. Despite this, we are seeing not just quantitative, but qualitative, advances.

From Crunching Numbers to Recognizing Patterns, Learning, and Developing New Strategies

The extraordinary growth in hardware capability has enabled software to make huge progress. Computers now are not just number crunchers. Attached to a camera, they recognize patterns; they observe similarities. They can tell whether an animal is a cat or a dog. They recognize faces. They can tell (as successfully as dermatologists) whether a skin lesion is malignant.

In 2002, the U.S. Defense Advanced Research Projects Agency announced a contest for a self-driving vehicle. The results, in 2004, were dismal. Not one of the 15 entrants made it one-tenth of the way into the course. Yet today, autonomous, self-driving vehicles can tell whether a darker area is a shadow across the road, an object in the road, or a hole in the road; and they can navigate appropriately. Waymo, Google's self-driving cars, have logged 20 million miles on public roads.

Results to date suggest that AI chauffeurs would sharply reduce the number of traffic accidents and deaths. This would allow hundreds of billions of dollars (just in the United States) spent on vehicle insurance to be diverted to more productive endeavors. It might also encourage more efficient design and use of vehicles.

Significantly, computers can learn. The most famous chess match in history took place in 1996 between the great Russian chess master (and courageous pro-democracy activist) Gary Kasparov and Deep Blue, an IBM computer. Deep Blue did not, could not, assess every possible sequence of moves. (There are infinite possibilities.) Rather, it was programmed with general strategies – but also the ability to override and change those strategies that did not work. It got better as it played more games.

In an even more complex game, in 2016, Google's DeepMind beat the world's top Go player. A year later, AlphaGo Zero, developed along the lines of DeepMind, but programmed with only the rules of the game (and no strategies), developed its own strategies by playing games against itself and quickly became a top Go player. In 2017, Libratus beat the world's best poker players.

Of course, these are games. Yet the same principles have been applied to more consequential matters. In medicine alone, computer-assisted cardiac surgery promises to provide more successful outcomes than conventional cardiac surgery. In addition,

> Many in the medical community were frankly surprised by what deep learning could accomplish: studies that claim AI's ability to diagnose some types of skin cancer as well or perhaps even better than board-certified dermatologists; to identify specific heart-rhythm abnormalities like cardiologists, to interpret medical scans or pathology studies as well as senior, highly qualified radiologists…
>
> These skills predominantly involve pattern recognition, with machines learning those patterns after training on hundreds of thousands, and soon enough millions, of examples. Such systems have just gotten better and better, with the error rates for learning from text-, speech-, and image-based data dropping well below 5%, whizzing past the human threshold.[11]

Dr Topol adds that relegating such tasks to artificial intelligence would render more accurate diagnoses at lower costs and *could* enable doctors to spend more time with patients.[12]

Although their native language is arithmetic, computers have become fluent in other languages. They can translate among them and can provide intelligent answers to questions posed in them. Chatbots have been used not only in the verbal exchange of information in communications with customers, but even to provide psychological therapy (https://woebothealth.com).

AI has added value to broad areas of our society. We have (semi) smart appliances, smart buildings that minimize energy consumption while keeping occupants comfortable. At a level removed from the consumer, we have smart autonomous tractors that monitor soil conditions. We have increasingly sophisticated industrial robots that run, wheel, jump, avoiding obstacles while lifting heavy loads (and the right ones).

From chauffeur to counselor, from tree removal to translator, from industrial designer to medical diagnostician, there seems to be little AI cannot do. Is there anything inherently beyond the potential scope of AI?

Creativity

A theme running through Western religions concerns the position of humans as the pinnacle of Creation. We live on the initial Creation in the universe and at its center. The sun, the stars, the planets, revolve around us. We are the only lives created with souls, in God's image. We are unique in our intelligence and our intrinsic value.

Over the past half millennium, science and technology have cast increasing doubt on this narrative. We live far from the center of the universe. The Earth was not created first, but was born later, and from a rather ordinary star. The stars do not revolve around us. We and monkeys (and other species) have common ancestors. We are their cousins. And there were other human varietals before *homo sapiens*, several of which lived at the same time as our direct ancestors.

Moreover, evolution developed incrementally. There may not have been a non-arbitrary first human, a first animal with a soul. (But if that is the case, no animal could have a soul – unless souls are incremental.) Species other than our own show intelligence, empathy, and even a sense of fairness.[13] As if these affronts to our special dignity were not enough, we now have AI challenging one of the most precious aspects of our uniqueness: our creativity.

The arts may be the most direct expression of unique creativity. One cannot find cave drawings, poetry, or musical compositions by other species. Yet computers have generated poetry, and also music, and even art. An AI program from Japan produced a short novel that came close to winning a literary prize.[14] A computer-generated portrait, *Edmond de Belamy*, sold at Christies for $432,000. But is it art? Does it deserve to be called a new creation?

> Up close, however, the paintwork becomes a grid of mechanical-looking dots, the man's face a golden blur with black holes for eyes. Look into those eyes. They show no sign of feeling or life… Art is a way in which human consciousness expresses itself… It has no existence outside the human passion to communicate… For a robot to make art, it would need an autonomous mind that was emotional as well as rational.[15]

As a more general defense of the uniqueness of our species, along similar lines:

> The problem with AI is that it can't create anything... The problem of equating creativity to math is that math isn't creative. To be creative implies to develop a new pattern of thought – something that no one has seen before... Creativity also implies developing a different perspective, which is essentially defining a different sort of data set (if you insist on the mathematical point of view). An AI is limited to the data you provide. It can't create its own data; it can only create variations of existing data – the data from which it learned.[16]

This raises a difficult question: *Just what is creativity*? Might Ecclesiastes have been right in saying that there is nothing new under the sun? Might it be that most human creativity does not venture far from pre-existing paradigms (the opening of Mozart's string quartet K. 465 an exception that proves the rule)? Might what we take as creativity be no more than recombinant Denatured Neolithic Artifact?

AI challenges our understanding of creativity. Was Deep Blue, abandoning chess strategies that failed and developing others, creative? Was AlphaGo Zero, not programmed with any Go strategies, but developing new strategies, perhaps ones that had never been seen before, in the course of playing against itself, creative? Might we translate this apparent AI learning from games into the arts, exposing a computer to great works of art, literature, music, and instructing it to create – and then grading its creations by a panel of expert humans, the pattern of the feedback from those grades then being used to improve subsequent rounds of creations? Might there be a Turing test to see if humans could distinguish between human-generated works of art and AI art?

In the last few decades, research in areas such as neuroscience and behavioral economics allowed scientists to hack humans, and in particular, to gain a much better understanding of how humans make decisions... Vaunted 'human intuition' is in reality 'pattern recognition.'[17]

Economic Ramifications

Economic growth is the product of productivity growth multiplied by the population (at least, labor market) growth. (There may be feedback between the two. A low level of productivity may limit the population, for not enough is produced to sustain more people. This had been the case for millennia. Higher productivity may limit the population for a different reason – the provision of a social safety net removing the incentive to have large families as insurance, so that social safety nets could provide a solution to excessive population growth.)

The primary driver of productivity has been new tools or technology, the impact of such tools being a product of how widely they are used multiplied by the average amount they increase individual productivity. Innovation, making capital more valuable, may have accounted for 80% of U.S. economic growth.[18]

The more potent the innovation, the wider its impact, and the more it increases productivity, the greater its contribution to economic growth. Other things being equal, the more it contributes to economic growth, the faster the economy will grow, to the potential benefit of all.

AI is one of the most potent innovations in history. It is a broadly applicable general-purpose technology. There are few, if any, sectors of the economy that lie beyond its productivity-enhancing influence. One might reasonably expect AI to be a powerful engine of economic growth.

This optimism is exemplified in an industry report by Accenture.

Accenture research on the impact of AI in 12 developed countries reveals that AI could double annual economic growth rates in 2035 by changing the nature of work and creating a new relationship between man and machine. The impact of AI technologies on business is projected to increase labor productivity by up to 40 percent and enable people to make more efficient use of their time.[19]

Similar optimism is reflected in the estimate of Forrester Research that AI will create 15 million jobs in the United States over the next decade.[20]

Such optimism, though it is founded on ample theoretic support, may well create a dissonance. For computers have been around for more than half a century, and AI has been around for decades. Yet the industrialized countries, at the forefront of these technologies, have experienced declines in productivity growth and concomitant economic growth. If AI is so potent an innovation, how might one explain this?

A satisfactory account may encompass several components:

One component, largely exogenous, involves the five 'Ds': Diversion, Debt, Disparity, Devaluation, and Deceleration. Each of these impediments to the economic growth of industrialized countries has grown in recent decades and any one may overwhelm the economic benefits provided so far by AI.

Diversion: In theory, trade should benefit everyone. People would not trade unless they benefit from the transactions. In practice, however, international trade has created a large share of losers, primarily labor in manufacturing that can be diverted from industrialized countries to lower labor-cost developing countries. This diversion of manufacturing, hollowing out industry in the developed countries, has taken an economic and productivity toll on those countries.

Debt: Debt has grown considerably over this period, and encouraged by central bank policies of low-to-negative real interest rates, is likely to continue to grow. However, excessive levels of debt are an economic drag and a destabilizing force, rather than a stimulus. Debt service transfers money from those who spend their income and borrow to those who do not and save, reducing the velocity of money, total spending, and GDP. At a micro-level, counter-party risk is exacerbated by excessive debt; and at a macro-level, there is less room to raise interest rates to fight inflation, for the economic costs of higher interest rates would be severe.

Many consumers have been spending their entire earnings plus borrowings. As they near the limit of their borrowing, they have no choice but to cut back on spending, to the detriment of economic growth. Similar considerations may apply to corporate and asymmetric national debt.

Disparity: Economic disparity, reversing the salubrious trend of *les Trente Glorieuses* and the Great Compression, has also been widening.[21] The potentially corrosive effects of this widening have been carefully examined, and the issue has attracted attention from economists and a few enlightened politicians.[22]–,[23], [24] The immediate economic impact of this may be limited, as increased spending due to the wealth effect may offset diminished spending by the rest.[25] In the longer term, however, excessive disparity may account for the consistent underperformance of those countries characterized by extractive, versus pluralistic, economic institutions.[26] As inequality grows in the industrialized countries, decreasing demand from the great majority may outpace increasing demand from the wealthy elites.

Devaluation: Standard measures of productivity and GDP may devalue the contribution of AI. A robotic surgeon might be more precise than any human surgeon could be. It could work seven days a week, 24 hours per day. It would yield a net improvement in life expectancy and quality. Yet because it is not paid as a surgeon, it would diminish GDP.

Deceleration: This considers the possibility that the initial burst of productivity-enhancing innovation, initially driven by the steam engine, has finally run out of steam. The most recent major innovations to penetrate all areas of the economy may be the commercialization of electricity (powered by steam) and the internal combustion engine (a straightforward extension of the steam engine). These innovations have matured, contributing a decreasing impetus to productivity and economic growth.[27]

This is not to deny that many valuable innovations have been made over the past decades. It is just that these have not have had the far-reaching economic impact of the earlier innovations. While positive, their impetus may have been insufficient to replace the declining impetus of earlier innovations.

Even if AI contributed significantly to productivity and economic growth, that might be insufficient to overcome the negative effects of these Ds.

In addition to the exogenous Ds, there may be an endogenous component that has to do with the nature of exponential growth. Something that grows exponentially doubles every N year (s). Its growth in any N-year period will exceed cumulative growth before that period. These things may start out small. But no matter how tiny is their start, after a sufficient time their growth rate and their size will exceed any limit. (Even the steam engine started slowly.)

Might this be what is happening with AI? Might it be that AI is still too small to be noticeable in the face of other, larger, macro-economic factors, but it will to continue to accelerate, thanks to continued growth and more widespread and effective dispersion, to the point that it dwarfs these other factors? This gives rise to several important questions.

Eventually, the acceleration of AI growth will have to subside. But how much will it subside? Will it converge to zero? Will it oscillate as new significant applications are developed and subsequently mature? Will it find an equilibrium at some negative level, at which point AI growth itself will slow? Separately, how large will AI grow by the time of that subsidence? What are the factors that will eventually impede the current extrapolations and be the causes of that subsidence?

These are significant, if difficult, questions that have received insufficient attention. However, while indefinite extrapolation is always problematic, there is a burden of proof, or at least reason, on skeptics to provide careful analysis of impediments to continued future acceleration of this sector.

危 – Danger

AI may be one of the few major innovations – perhaps the only one in history – whose danger stems not from malicious intent, but from the very nature of the innovation itself. The danger is manifest in two inter-related questions:

1   Is there anything of economic value that a person could do that AI could not (perhaps at some point in the future) do at least as well?
2   Is there any job or profession in any sector of the economy in which humans could not be profitably replaced by AI?

Optimism with respect to the future of AI – partly driven by a history of underestimating what AI could achieve – implies pessimism with respect to these questions. While we justly celebrate the Industrial Revolution, it produced widespread devastation for generations. Charles Dickens was not exaggerating in his dismal portrayal of lives.

> Real blue-collar wages in Britain were almost halved between 1755 and 1802... This period of intense technological progress in the United Kingdom was also an era of intense deprivation and very difficult living conditions. The economic historian Robert Fogel showed that boys in England during this period were significantly undernourished compared even to slaves in the US South.[28]

The new period of 'adjustment' could be even more far-reaching, wiping out all jobs for all time.

The Economic Challenge

It may be instructive to start by considering a natural, but dystopian, extreme. One would then inquire how to avoid such an extreme.

The extreme arises from continued near-exponential AI progress in an environment of *laissez faire*, non-interference. This would increase the profitability of replacing any human labor by AI. Some jobs – bank tellers, drivers, highly-paid financial traders (Goldman Sachs has replaced its 600 traders with two traders plus AI)[29] – would be replaced sooner; others only after AI had advanced considerably from its present state. But no human labor would be exempt. All income would eventually flow only to those who hold capital.

This could be regarded as a plausible default scenario. Nothing needs to change from present trends to get there. It may be more difficult to prevent such a scenario because it is likely to occur slowly, like global warming, enabling beneficiaries with capital to deny any deleterious effects.

However, the fact that this default is natural does not make it palatable. In the extreme case of economic rationality, the great majority, who do not have access to adequate capital, would either starve or be reduced to subsistence farming and hunter-gathering (if they could access the land).

Even slavery and serfdom would not provide a means of survival, as AI and robotics could furnish more efficient slaves. It would be (on conventional theory) economically irrational to employ, or even enslave, humans. Better to let them starve.

Few would find such a scenario acceptable. However, it is dangerously easy to underestimate how readily something like this could unfold. Admittedly, there is cogent reason to reject such a scenario. However, history has often trumped rationality – and morality. In keeping with our past and the notion that wealth is a sign of grace (or at least of creative industry) and that poverty is a sign of moral weakness or sloth, there is a palpable danger that the history of policies that favor and entrench established wealth will trump such reason and will enable, even justify, such a dismal scenario.

One can already encounter arguments that favor regressive economic policies. Moreover, there is a vicious circle in which such policies strengthen the political leverage of the plutocrats who advocate them.

A common argument against spending on social services, often heard in the United States, is that the poor are unworthy and deserve their poverty. They are lazy and lack a robust work ethic.[30] A related argument against strongly progressive taxes, also commonly heard in the United States, is that they sap the incentive of the most productive and industrious people, rewarding (via lower tax rates) the lazy poor.[31] An argument against inheritance taxes, again common in the United States, is that those who have worked hard to earn money have the right to dispense it as they see fit – and that this is a right that extends to those who have not worked hard, but have merely inherited their fortune.[32] Moreover, it is the right to pass on that wealth, free of taxes, that keeps them from wastefully spending it.

None of these arguments are persuasive. And in an economy dominated by AI, they lose every shred of cogency. Despite this, wealth has a disproportionate political influence in all countries, and it is conceivable that it would use that influence to effectively oppose significant economic redistribution.

Especially in countries that already have plutocratic tendencies, that influence may be powerful.

Extreme economic disparity, an effect of all income flowing to capital, coupled with wealth being entirely determined by inheritance (as all labor is more efficiently performed by AI), would create *une mauvaise époque*. It may be critical to address this proactively, before plutocratic institutions harden.

Addressing this requires the acknowledgement that what is natural is often not benign. Just as we wear clothes and build houses to shelter ourselves from the raw forces of nature, it may be necessary to build institutions to shelter ourselves from the raw forces of economics. Moreover, as these raw forces change, it may be important to adjust the institutions.

The development of AI may represent a significant change in raw economic forces. For as AI advances, to paraphrase the song from *Annie Get Your Gun*, 'Anything humans can do, AI can do better.' The increased efficiency of AI will provide greater surplus. But the independence of that surplus on human effort or merit will provide a serious challenge with respect to its distribution.

Merit and incentive are no longer relevant. Rights of inheritance, as they become further removed from any merit that is the source of that inheritance, become more tenuous. Arguments for taxing any income – which beneficiaries may have done nothing to earn – become more difficult to evade. It is an increasing strain to find reasons that would justify inequality.

However, this must be weighed against the (current) political reality of the disproportionate influence of established wealth. It may be that minimal decency requires a social safety net that includes quality housing, nutrition, medical care, education, and disposable income. Yet governments, even those capable of providing for citizens, have often failed the decency test. Political choices will be crucial, and it will be important to emphasize the irrelevance of those popular economic arguments (that often appear to inform political decisions) that disparity generates incentives and provides just rewards.

The Existential Challenge

This goes beyond the economics. If *Anything humans can do, AI can do better*, what is the point in trying? What constitutes worthwhile purpose?

That leads to questions as to what makes life meaningful. Wise government action, highly progressive tax codes coupled with generous and secure social safety nets, may remove economic considerations from this question. All economically meaningful labor would be performed more efficiently by AI, and government policies would assure adequate means for a secure comfortable life for all. That could lead worshippers at the Church of Mammon to reassess their priorities.

Relieved of the economic dictum that every good has a monetary equivalent, people may develop a more profound understanding of and appreciation for values that are often ignored or taken for granted: for health, for deep and warm personal relationships and the concern of others, for one's own life (and those of others), for dignity and respect, for one's growth as a person, for a secure and sustainable environment, for societies that enable and encourage all to develop their potential to the fullest.

As two recent Nobel Laureates write:

> *The focus on income alone is not just a convenient shortcut. It is a distorting lens that often has led the smartest economists down the wrong path, policy makers to the wrong decisions, and all too many of us to the wrong obsessions… Restoring human dignity to its central place, we argue in this book, sets off a profound rethinking of economic priorities…*[33]

Virtue – 德

As the difference between the consequences of winning and those of losing grows, so does the incentive to sacrifice morality in order to win. Survival is a powerful incentive. People reduced to struggling to survive, often at the expense of others, may be hard-pressed to develop virtue. As more people sacrifice morality, that sacrifice seems more normal, less egregious in the eyes of others, triggering a downhill spiral that is not easily reversed.

If, however, economic survival with some degree of comfort is assured, the quaint notion that virtue is its own reward, an epitome of conventional economic irrationality, may become fashionable. (If one's motive for doing good deeds is some reward – money, prestige, heaven, karma – then is one really being virtuous?) That could foster healthier priorities and proclivities and heighten expectations, which themselves could lead to further improvement. This would lead to the opposite of the dystopian scenario created by economic nature taking its course.

However, it may be that the action – or inaction – of government will play a role in the quality of society and the lives of its people. For centuries, national success has required robust, as well as pluralistic, central government.[34] Societies without these have historically been more primitive, more Hobbesian, less virtuous.

Just as governments contribute to the literacy and numeracy of their citizens, important positive externalities, by providing public education, they may contribute to the virtue of their citizens, another important positive externality, by providing a sufficient cushion against privation, by limiting the negative consequences of losing. In an economy driven by AI, that may be easier thanks to greater surplus. It may also be essential.

In effect, AI increases the danger but also elevates the opportunity. The dystopian danger, an effect of labor losing all its economic value, may be evident. Yet there is opportunity as well. Limiting concerns of survival and even comfort, while also limiting the benefits of the accumulation of endless wealth may free people to develop a broader and more appropriate notion of value, to live better lives, to create more enduring value.

Notes

1 Kron, G. (2013) 'Fleshing Out the Demography of Etruria,' in *The Etruscan World*, ed. J. M. Turfa, Routledge: London.

2 Morris, I. (2010) *Why the West Rules – for Now*, Farrar, Strauss, Giroux: New York, p. 629.

3 Pinker, S. (2011) *The Better Angels of our Nature*, Viking Press: New York.

4 Whitmore, D. (2018) 'Economic Myths #9 – Social Safety Nets,' The Ludwig von Mises Center for Property and Freedom: Auburn, AL.

5 Kendall-Taylor, A., Frantz, E., and Wright, J. (2020) 'The Digital Dictators: How Technology Strengthens Autocracy,' *Foreign Affairs*, 99:1, www.foreignaffairs. com/articles/china/2020-02-06/digital-dictators.

6 Solow, R. (1957) 'Technological Change and the Aggregate Production Function,' *Review of Economic Statistics*, 39, pp. 312–20, www.jstor.org/stable/1926047.

7 'Productivity and Progress', *Monthly Labor Review*, U.S. Bureau of Labor Statistics (September 5, 2017), www.bls.gov/opub/mlr/2017/book-review/produ ctivity-and-progress.htm.

8 Morris, I. (2010) *Why the West Rules – for Now*, Farrar, Strauss, Giroux: New York.

9 Brynjolfsson, E. and McAfee, A. (2014) *The Second Machine Age*, W. W. Norton: New York, pp. 49–50.

10 Mandel, M. and Swanson, B. (2017) 'The Coming Productivity Boom: Transforming the Physical Economy with Information,' www.techceocouncil.org/clientuploads/ reports/TCC%20Productivity%20Boom%20FINAL.pdf.

11 Topol, E. (2019) 'Interview,' *Life Extension* 25:12, p. 77.

12 Topol, E. (2019) *Deep Medicine: How Artificial Intelligence Can Make Healthcare Human Again*, Hachette Book Group: New York.

13 De Waal, F. (1996) *Good Natured: The Origins of Right and Wrong in Humans and Other Animals*, Harvard University Press: Cambridge, MA; Safin, C. (2015) *Beyond Words: What Animals Think and Feel*, Henry Holt and Company: New York.

14 Olewitz, C. (2016) 'A Japanese Program Just Wrote a Short Novel, and It Almost Won a Literary Prize,' *Data Trends*, March 23, 2016, www.digitaltrends.com/cool-tech/japanese-ai-writes-novel-passes-first-round-nationanl-literary-prize/.

15 Jones, J. (2018) 'A Portrait Created by AI Just Sold for $432,000. But Is It Really Art?' *The Guardian*, October 26, www.theguardian.com/artanddesign/shortcuts/ 2018/oct/26/call-that-art-can-a-computer-be-a-painter.

16 Mueller, J. P. and Massaron, L. (2018) *Artificial Intelligence for Dummies*, John Wiley: New York, pp. 226–7.

17 Harari, Y. N. (2018) *21 Lessons for the 21st Century*, Random House: New York, pp. 20–21.

18 Solow, R. (1957) 'Technological Change and the Aggregate Production Function,' *Review of Economic Statistics* 39, pp. 312–20, www.jstor.org/stable/1926047.

19 Purdy, M. and Daugherty, P. (September 2016) *Why Artificial Intelligence Is the Future of Growth*, Accenture, https://dl.icdst.org/pdfs/files2/2aea5d87070f0116f 8aaa9f545530e47.pdf.

20 Rouhiainen, L. (2019) *Artificial Intelligence*, Createspace Independent Publishing Platform, p. 109.

21 Piketty, T. (2013) *Capital in the Twenty-First Century*, Harvard University Press: Cambridge, MA & London.

22　Kuttner, R. (2018) *Can Democracy Survive Global Capitalism?* W. W. Norton: New York.

23　Acemoglu, D. and Robinson, J. (2012) *Why Nations Fail*, Crown Business: New York.

24　Sitaraman, G. (2017) *The Crisis of the Middle-Class Constitution*, Alfred A. Knopf: New York.

25　Zandi, M. (2017) 'What Does Rising Inequality Mean for the Macroeconomy?' in *After Piketty*, ed. H. Boushey, J. B. Delong, and M. Steinbaum, Harvard University Press: Cambridge, MA & London.

26　Acemoglu, D. and Robinson, J. (2012) *Why Nations Fail*, Crown Business: New York.

27　Gordon, R. (2016) *The Decline and Fall of American Growth*, Princeton University Press: Princeton, NJ.

28　Morris, I. (2010) *Why the West Rules – for Now*, Farrar, Strauss, Giroux: New York, pp. 230–31.

29　A. Zhygalina (2018) 'Neural Networks in Trading: Goldman Sachs Has Fired 99% of Traders Replacing Them with Robots', Crypto News, 18 September, https://cryptonews.net/editorial/technology/neural-networks-in-trading-goldman-sachs-has-fired-99-of-traders-replacing-them-with-robots/.

30　Murray, C. (2013) Coming Apart: The State of White America 1960–2010, Crown Forum: New York.

31　Hagopian, K. (2011) 'The Inequity of the Progressive Income Tax,' The Hoover Institution, www.hoover.org/research/inequity-progressive-income-tax.

32　Nozick, R. (1974) *Anarchy, State, and Utopia*, Basic Books: New York.

33　Banerjee, A. and Duflo, E. (2019) *Good Economics for Hard Times*, Hachette: New York, p. 9, italics added.

34　Acemoglu, D. and Robinson, J. (2012) *Why Nations Fail*, Crown Business: New York.

8 Conclusion

Ruth Taplin

The old adage to be forewarned is to be forearmed is very true in relation to developments in AI which will move relentlessly forward in a new phase of the digital revolution that unites AI, IPR, Cyber Risk and Robotics. How AI is used and implemented will have the greatest ramifications for both positive and negative influences. One major point to take note of is that AI is not a living system, nor is it autonomous from human decisions. To think otherwise can only lead to those with unsavoury intentions blaming a technology for ill deeds rather than the humans behind the decision making.

This does not mean that humans will refrain from working hard to imbue AI robots and learning machines with ever more human features and personas.

At the Beijing Winter Olympics in 2022, robots were in evidence everywhere, preparing food, cocktails and cleaning with anti-virus spray to prevent COVID-19 among those attending. Unfortunately, China is currently experiencing a renewed wave of COVID-19 cases and deaths despite its AI robots learning how to spray disinfectant.[1]

In the ongoing tensions with India over border areas in which China has been building structures on Indian land, a move which is viewed as expansionist by India, China is deploying AI robots to maintain a constant presence on the borders without committing human soldiers to such inhospitable terrain. The disputed border lands have little practical value, but the dispute concerns which country will be seen as dominant on the border regions, especially Chinese-controlled Tibet.

Using its AI technology, China has deployed machine-gun-carrying robots and unmanned vehicles which can carry both weapons and supplies to Tibet and its borders with India. The vehicles include the Sharp Claw mounted with a light machine gun that can be operated wirelessly. The Mule-200 has also been sent to the border; this is an unmanned supply vehicle that can also carry weapons. Lynx all-terrain vehicles, which can carry weapons systems such as mortars, missile launchers, heavy machine guns and howitzers, are also in use.[2]

Unfortunately, the conflict continues on the Indian–Sino borders without abatement, with AI having little impact on geopolitical considerations.

As the war by Russia against Ukraine continues, the physical attacks are soon to be followed by predicted massive Russian cyberattacks that target

DOI: 10.4324/9780367857561-8

basic infrastructure such as electric smart grids. The 'Five Eyes' intelligence-sharing network which consists of the United States, the United Kingdom, Canada, Australia and New Zealand is finding evidence that the Russian government and domestic cybercrime hackers that they use in tandem when needed are planning massive attacks against Ukraine, but also all governments that support Ukraine globally.[3]

AI will undoubtedly be used, as explained in this book, to counter such Cyber Risk and attack.

Botnets, IoT and AI

AI is facilitating the ferocity and effective destructiveness of cyberattacks through the Internet of Things (IoT), defined in the author of this chapter's last book on *Managing Cyber Risk in the Financial Sector*, as interconnected infrastructure sensors embedded in objects, which collect and transmit data concerning their internal states or environment to their users in real time, where they are analysed or fed into models. Through its digital interconnections, the IoT provides fertile ground for invasive malware. As mentioned in the Cyber Risk book, the Mirai botnet attack in 2016 was massive.[4] It was a distributed denial-of-service (DDoS) cyberattack that consisted of hundreds of thousands compromised IoT devices that targeted Dyn, a Domain Name System (DNS). The launching of the malware attacks had been carried out through common devices such as video recorders and digital routers.

The largest DDoS attack ever recorded was in 2018. It targeted the GitHub software development platform and took the platform offline. Cyberattacks such as these have not stopped and with the Russian aggression, the situation will only worsen.

Ironically, AI machines, which can be used to good effect to stop such attacks, are the very medium that is allowing these attacks to expand through automation and the rapid processing of data patterns as noted in this book. The threat continues to increase with Bot-as-A-Service being used by cyber criminals to outsource their services.

To become more effective at using AI-driven cyberattacks, governments such as Russia and China, as mentioned, work in tandem with cyber criminals to launch more ferocious and destructive malware attacks such as DDoS, along with ransomware, to cripple other governments geopolitically, and in cyberspace ransack companies for political or pecuniary gain.

AI Can Be Used to Mitigate Cyberattacks

As mentioned in this book, AI-generated algorithms can be used in a number of ways to mitigate botnet cyberattacks. As explained, one of the strengths of AI machine learning is that it can work in real time. This allows the person operating the computer to analyse patterns and the behaviour of the bot in real time – to understand the direction of the bot traffic in order to combat it effectively. This can work by cross-checking all visitors and checking their signature against the person's computer database.

There are a number of bot detection tools that are emerging to expose a bot's true identity. Unusual header information or web requests can give a clue to malware attempts which can then be blocked immediately.

It is important, as with all cyberattacks, to cooperate with other companies and governments to share information on stopping them. One cybersecurity firm named HUMAN Security has been defeating botnets through aggressive cooperative approaches with industry and law enforcement agencies. They synthesise behavioural pattern data with a real-time decision making engine that combines machine learning (ML) with technical evidence to ensure that only genuine human interactions are occurring. Companies like HUMAN Security, using AI machine learning reverse engineers, uncover and disrupt threats by bots to cybersecurity, marketing and advertising. With Roku and Google, for example, they stopped PARETO, which was the most sophisticated Connected TV to date.[5]

Future of Regulation for AI

Part of the problem of destructive uses of AI is that regulation at a governmental level is lacking globally. Europe is far ahead of most countries in terms of regulations and also leads the United States which is lagging behind in terms of regulation.

Margrethe Vestager, the digital chief of the European Commission (the EU governing body), stated that AI systems which are considered a threat to the safety and rights of people in society would be banned through the rule of law.

Rules will be tightened on biometric use. Use of facial recognition by law enforcement agencies would also be subject to the law. Harsh penalties are being proposed for companies that abuse biometrics and AI facial recognition systems with a variety of fines that can be percentages of global turnover.

The EU is attempting to develop new global standards for the use of AI and all its applications.

Contrary to a number of AI companies which believe that the proposed new EU regulation will stifle innovation and could be repressive, Vestager argued that most AI is considered low-risk and will not be interfered with; regulation instead will concentrate on serious breaches of human rights such as the use of subliminal control that undermines people's freedom of thought. Chatbots used in customer services, for example, would not be subject to bans as long as people are made aware they are speaking to an AI machine.

AI used for coercive purposes – such as the extensive Russian use of bots during the 2016 US Presidential election which pretended to be real people but were spreading disinformation – would be banned; but biometrics used to deter terrorist acts, such as the murder of the UK MP Sir David Amess by a student of Somalian origin, would not be banned.

The EU is working on fine-tuning what are acceptable and unacceptable uses of AI within Europe and what remedies should be used for destructive, undemocratic uses of AI. It will also add clarity to what has been discussed in this book concerning AI robots being granted patents and falsely being considered to be independent, autonomous inventors.

The EU is also drafting a related internet law, termed the Digital Services Act (DSA), which will ensure democratic control over the use of algorithms, ruling out purposeful disinformation, a ban on advertisements targeted at minors and support for the democratic rights of EU citizens. The EU is also seeking to make this a global standard of digital internet use.[6]

Future Positive Applications of AI

Telemedicine and Telecare

Telemedicine has been a positive digital method to save lives by treating patients remotely for decades. The Editor and author of four chapters in this book has seen this working to good effect in Australia which has very remote areas with a sparse population. Medical expertise resides in the urban areas which are mainly on the coasts, a long distance from the interior. Through telemedicine platforms, patients and local practitioners can be connected to medical experts via computer screens and monitored through telemedicine platforms.

In the UK, AI machines are enabling telemedicine and social care using sensors. In Dorset in 2021, there was a three-month trial to monitor and treat a hundred people who required social care. Given the UK crisis in social care and limited numbers of care homes, Lilli, a UK-based technology company, created the AI sensor-based devices that monitor movement, temperature and the use of particular appliances such as kettles.

Monitoring is based on the AI sensors, assessing how many times, for example, a patient just released from hospital opens their refrigerator, puts on the kettle or needs to make toilet visits. Such information will add to data on whether the person has developed a urinary infection or is able to feed themselves. Changes in behavioural patterns detected by the AI sensors can reflect if there are health problems or if patients are coping at home. Data are encrypted and only the organisation treating the patient will have access to the patient's personal data. The Lilli system will only install their AI systems with written consent from the patient or those authorised to sign on their behalf.

Telecare can both reduce the costs of social care and make it easier for those who have mobility issues. There could be issues of loneliness, as some patients may rely on social care visits, but these issues are often dealt with through social gatherings organised by the local council or charities. Telecare through AI sensors would be used mainly to monitor patients' health conditions.

Trials by the Lilli system are still ongoing as this is an AI sensor for the future of social care.[7]

AI: Military Applications Only?

Like most new technologies, AI was originally developed for military and defence purposes. AI continues to be primarily a military technology that carries out mundane tasks, such as those outlined in this book, to make data analysis more rapid and expand computer vision. This is why issues of cybersecurity are so integrally tied to the military and geopolitics as the more advanced the cyberattack, the more AI is needed to mitigate attacks against the military and governments in addition to companies. Yet AI is a dual technology that has civilian, private company uses as well as military applications according to the humans using it. In fact, the inventors of new AI technology are no longer those mainly connected to the military such as the US Pentagon; instead, AI inventions can be the preserve of tech giants which means that AI developments are not necessarily guided by state actors. An example of this dual technology dilemma occurred under the Obama Administration when Google ended its contract with the Pentagon and built an AI laboratory in Beijing, China. In 2017, President Xi Jinping added to the Chinese Communist Party's constitution the requirement that all research carried out in China must be shared with the People's Liberation Army. As mentioned in this book, China has been working towards advancing in every area of technology to further its ambitions to become a world leader.[8]

If those developing AI are supporting political regimes that have authoritarian agendas, the question will be raised as to whether it is the tech giants that are a danger to democracy, rather than AI with the 'killer robot' scenarios in which robots autonomously take on humanity with evil intention. Humans cannot abandon responsibility for their actions as they have in the past through the genocidal massacres which litter history or another Holocaust. This is why assigning inventorship and patent rights to the robot DABUS may seem innocuous, but in political reality can present the greatest danger of irresponsibility and lack of accountability by humans anywhere in the world. Perhaps this is the real reason why the United States Congress and Senate and the European Commission have been drafting regulatory laws to rein in the tech giants who are most often the developers of AI and constraining 'open access' or 'open source' policies.

The Future of AI Regulation

The future problems of reining in tech giants and the subsequent unregulated use of AI are complex and numerous. Tech giants are fully behind 'open access' and ' open source' ways of ultimately benefitting themselves, while undermining inventors' and creators' rights to protection of IPR or nation states' rights to security. But then they rail against allowing their users to upload their own security applications to protect themselves from installation of malware through tech giants' platforms and pernicious disinformation by state actors.

A recent example of this conundrum was reflected in legislation that were submitted to the US Congress in the early part of 2022. One of the two pieces of legislation was the Open App Markets Act proposed by Senators Richard Blumenthal (Democrat, Connecticut), Marsha Blackburn (Republican, Tennessee) and Amy Klobuchar (Democrat, Minnesota). The idea behind it was antitrust: to "promote competition and reduce gatekeeper power in the app economy, increase choice, improve quality and reduce costs for consumers".

The second piece of legislation, sponsored by Amy Klobuchar and Chuck Grassley (Republican, Iowa) was the American Innovation and Choice Online Act. This would prohibit tech platforms from "favoring their own products or services, disadvantaging rivals, or discriminating among businesses that use their platforms in a manner that would materially harm competition on the platform". This would allow smartphone users to allow competing Application stores onto their mobile phones. However, reducing the gatekeeper power of the tech giants and mobile companies through allowing unknown third-party applications to be uploaded by their users can breach user security and even national security.

These two pieces of proposed legislation cannot legally force the tech giants such as Amazon, Google, Apple or Microsoft to add unscreened applications to their products. But, because of the nature of the competition antitrust element in the proposed legislation, coupled with the tech giants' own open-access and open-source policies, they would be required to expend great amounts of time, money and expertise to screen all third-party application uploads. This would need to be done to protect both their own cybersecurity and national security.[9] Additionally, there could be breaches of Intellectual Property Rights as many of the uploaded applications could be infringing such rights without understanding they are doing so.

Artificial Intelligence and Machine Learning are the newest and most effective digital technologies that can and will be used to protect the environment, mitigate climate change and cyberattack, protect Intellectual Property Rights, develop robots that can assist humans with mundane and repetitive tasks, serve the military and fulfil a multitude of other applications while continuing to be non-living systems that are automated tools of humans.

The debate continues concerning the sentience of AI. Some creators of robots still think their creations are capable of independent thought. They are often dismissed as wishing this was the case, or seeing the situation based on science fiction rather than reality. These sentiments are well explained by Sam Altman, the CEO of OpenAI, a research lab in San Francisco that created GPT-3 (Generative Pre-trained Transformer 3) which is based on a neural network, as explained in this book. GPT-3 can write computer programmes, translate languages and summarise emails.

Yet, these capabilities are largely erratic or hit-and-miss in nature. For example, if asked to replicate ten speeches by a well-known person, GPT-3 may be able to do this five out of ten times, at best.

When challenged, Altman admits that AI is not the same as human intelligence, stating that it is like alien intelligence. He believes that AI developers are on a path to building a human brain and one that can experience sentience.

However, Altman admits that maybe these thoughts derive from excitement about the possibilities, and that some AI researchers cannot differentiate between reality and science fiction.[10]

There are many other AI developers who express varying views, being completely confident that AI creations by humans are sentient, can create patentable products independently of their human creators, and will be able to replace all human endeavour.

As the debate continues, it is always safest to err on the side of objective reality.

Notes

1 'Shanghai Lockdown: Whole Communities Relocated in Anti-COVID Drive', Robin Brant, BBC News Shanghai (21 April 2022), www.bbc.com/news/world-asia-china-61160210.
2 'China Replaces Soldiers with Machinegun-Carrying Robots in Tibet', Chris Pleasance, *Daily Mail* (29 December 2021), www.dailymail.co.uk/news/article-10352255/China-replaces-soldiers-machinegun-carrying-robots-Tibet.html.
3 'Russia-Ukraine War Latest: Ukraine Offers Unconditional Talks on Mariupol as West Warns of Russian Cyber-Attacks – Live', Samantha Lock, *The Guardian* (21 April 2022), https://digismak.com/russia-ukraine-war-latest-ukraine-offers-unconditional-talks-on-mariupol-as-west-warns-of-russian-cyber-attacks-live-ukraine/.
4 Ruth Taplin, *Managing Cyber Risk in the Financial Sector: Lessons from Asia, Europe and the USA* (Abingdon: Routledge, 2016).
5 'When Botnets Attack', Chuck Brooks, *Forbes* (22 April 2022), www.forbes.com/sites/chuckbrooks/2022/04/22/when-botnets-attack/?sh=72c0daef44df.
6 'EU Artificial Intelligence Rules Will Ban "Unacceptable" Use', BBC News Online (21 April 2022), www.bbc.com/news/technology-56830779.
7 'Sensors and AI to Monitor Dorset Social Care Patients', Chris Baraniuk, BBC News Online (27 August 2021), www.bbc.com/news/technology-58317106.
8 'Good for Google, Bad for America', Peter Thiel, *The New York Times* (1 August 2019), www.nytimes.com/2019/08/01/opinion/peter-thiel-google.html.
9 'New Antitrust Legislation Could Open the Door to Cybersecurity Problems', Guest Commentary by Chuck Brooks, Barron's Ideas (1 February 2022), www.barrons.com/articles/new-antitrust-legislation-could-open-the-door-to-cybersecurity-problems-51643664301.
10 'AI Is Not Sentient. Why Do People Say It Is?', Cade Metz, *The New York Times* (5 August 2022), www.nytimes.com/2022/08/05/technology/ai-sentient-google.html.

Index

For Product Safety Concerns and Information please contact our EU
representative GPSR@taylorandfrancis.com
Taylor & Francis Verlag GmbH, Kaufingerstraße 24, 80331 München, Germany

www.ingramcontent.com/pod-product-compliance
Lightning Source LLC
Chambersburg PA
CBHW060316220326
41598CB00027B/4342